Elementary Mathematics from an algorithmic standpoint

by Arthur Engel

English translation by F.R. Watson

Keele Mathematical Education Publications
University of Keele, Staffordshire, U.K.

1st edition, 1984 ISBN 0 947747 00 1

Translated from the original German edition of 1977, Klett Studienbücher series, published by Ernst Klett Verlag, Stuttgart:-

"Elementarmathematik vom algorithmischen Standpunkt" (ISBN 3-12-983340-4)

Figures and diagrams reproduced, by permission of Ernst Klett Verlag, from originals by G. Wustmann, Stuttgart.

Printed by Alsager Printing Co. Ltd.
Published by KMEP, University of Keele, Staffs, U.K.

Foreword

All who are interested in the development of the secondary curriculum for mathematically able pupils should study the writings of Arthur Engel. His work on geometry and on probability and statistics is outstanding, and this volume on algorithms confirms him as one of the most original teachers of our time.

Engel shows at once the unity and the continuous development of mathematics. He can produce instructive examples in an appropriate didactical sequence, and sharpen the argument until some apparently simple method acquires unsuspected power; then looking back we see that he has been giving us a history lesson as well.

The reader who is not used to his style may occasionally be baffled by terse, epigrammatic lines of proof - but he always has a reason for the line he takes and it is more than worth the effort to master the ideas from his perspective.

Those who are daunted by the task of reading a textbook in German may thank Mr F. R. (Joe) Watson for preparing this English translation. I hope that readers will not only study the text, but will go further and use this book as a source of problems for practical investigations on microcomputers. This may not be easy - but the rewards are great.

Trevor Fletcher

PREFACE

At the turn of the century Felix Klein launched a reform of mathematics instruction. The reform movement adopted the slogan "functional thinking". The function concept was to be the key concept from which the whole fabric would emerge. The widespread use of computers and pocket-calculators makes the time ripe for the next stage of reform, with the slogan 'algorithmic thinking'. The concept of algorithm should now serve as the key concept for mathematics in schools. We need to think out afresh the material of the school syllabus from the algorithmic standpoint. This book is intended to help in the process of rethinking. It deals with the heart of computer science - that is the construction and testing of algorithms. Moreover, it restricts itself preponderantly to the field of school mathematics.

Special emphasis is given to the many elementary and rapid algorithms for the transcendental functions which can be evaluated much more easily in this way than by the use of series.

The main emphasis of the book is on algorithms. But at the same time it is also a multi-faceted mathematics text which will broaden and deepen the reader's mathematical knowledge. As a by-product the reader gains experience in computer programming.

From this choice of theme it follows that the book is directed at three groups of readers

(a) Mathematics teachers in training in universities and teacher training colleges. The book developed out of a course on computer-oriented mathematics which has been regularly taught at the University of Frankfurt.

(b) Mathematics teachers who are investigating the possibility of integrating computers or calculators into mathematics teaching.

(c) Those following courses in informatics (computer science), algorithms or numerical mathematics in the upper secondary school.

It supplements the pupil's knowledge with a repetition and substantial deepening of the material. At the end of a school course such a review and overview is more sensible than superficial coverage of a totally new topic. A greatly-simplified version of this book was tried out successfully in the U.S.A.

The majority of the algorithms which appear can be carried out quite conveniently on a pocket calculator. Consequently the book can be studied successfully even if only a pocket calculator is available.

The book is very rich in content. The material cannot be covered in a single half-year. The major ideas are conveyed by suitable examples. The individual sections are almost independent of one another so that they can be omitted at will (especially those marked *). Although numerical algorithms preponderate it is possible by careful choice of the examples to provide a course which emphasizes non-numerical algorithms.

Despite some misgivings I have decided to use the BASIC language. It is easy to learn, it is the most widely used language in schools and it is available on microcomputers. Nevertheless, in the second half of the book an enhanced BASIC is used which would not be 'understood' by most microcomputers. The reader will need to translate the given programs into the particular dialect of his machine. Most algorithms are formulated in language-independent form. Anyone who uses another language such as Pascal-E will be little inconvenienced by the use of BASIC.

In the book decimal-point and decimal-comma are treated as interchangeable i.e. no distinction is made between 3,14 and 3.14. The numbers of the more difficult exercises are enclosed in square brackets. Interesting exercises are not specially indicated, for about one third of the 235 exercises fall in this category. Solutions of the exercises will be found at the end of the book.

I am particularly indebted to Herr Horst Sewerin. He has discovered a multitude of errors in the proof. Undoubtedly the reader will find others.

Frankfurt: 16th January 1977. Arthur Engel

TRANSLATOR'S PREFACE

Very few alterations have been made in the text. Updating or additions would have further delayed the appearance of this English edition. The intention is to make the original and stimulating material available to a wider audience and it is essentially a translation of the 1977 German edition, which attempts to reproduce the lucid and attractive style of the original. Much of the translation was carried out whilst on study leave at the Institut für Didaktik der Mathematik, Bielefeld, and I thank Prof. H. Bauersfeld for providing the opportunity for me to work there.

A few trivial misprints have been corrected (and doubtless others have been introduced!) but translator's notes have been kept to a minimum. These are indicated by the symbol † ; they are for the most part a consequence of the wider availability of more powerful BASIC and of graphics facilities in the past few years. Programs and figures have been reproduced unaltered (with very few exceptions); this means that some variable names correspond with their German equivalents - thus G (Gewinne) rather than W (wins), K (Kain) rather than C (Cain).

I am grateful to Herr Engel and to Klett Verlag for permitting the publication of this English edition. This project has met with a number of difficulties and delays which were in no way their responsibility, and I thank them for their patience and cooperation. My sincere thanks are also due to Gill Bailey, Mary Johnson, Angela Mason and Blair Brady for their valued help at various stages of the production, and to Dr T. J. Fletcher for his encouragement and for writing the Foreword.

F. R. Watson
Keele, Jan 1984

A selection of the BASIC programs given in this book has been adapted to run on the Acorn-BBC microcomputer. The programs are available on disc or cassette - enquiries should be addressed to KMEP, Dept. of Education, University of Keele, Staffordshire, ST5 5BG.

C O N T E N T S

page

1. Algorithms — 1
   - Algorithms and programs
   - The 3A+1 problem
   - Algorithmic definition of some functions
   - The maximum- and the minimum- function
   - Factorials and powers
   - The whole number function (INT)
   - Roots
   - Logarithms, random numbers and powers
   - Random number generators
   - Fibonacci series

2. Number theory — 43
   - Conversion from base 10 to base 'b' and vice versa
   - Euclidean algorithm
   - * Extension of Euclidean algorithm
   - Prime numbers
   - Periodic decimals
   - Continued fractions
   - * "Chinese prime numbers"

3. Geometry — 69
   - Archimedes' method for area under a parabola
   - Calculation of π
   - Algorithms for trigonometric functions
   - Method of Cusanus (π by inscribed and circumscribed polygons)
   - Area under a hyperbola
   - Speeding up convergence : the Romberg procedure
   - Lattice points in a circle (π by counting)
   - Leibnitz series for π
   - Monte-Carlo methods for determining π

4. Numerical mathematics — 109
   - Solving an equation
   - Maximum of a unimodal function
   - Numerical integration
   - Trapezium, mid-point and Simpson's rules
   - Romberg procedure
   - Proof of Simpson's and Hermite's rules
   - Differential equations
   - Growth, decay and oscillations
   - Numerical integration of differential equations
   - Simulation of dynamical processes
   - Harmonic series
   - Calculation of e to 250 places

5. Combinatorics and probability — 173
   - Program for Pascal's triangle
   - Frequency tables
   - Permutations
   - Probability problems

6. Simulation of random processes — 191
   - A random number generator
   - Simulation with a random number generator
   - Simulation without a random number generator

7. Sorting — 220

8. The 8 Queens problem — 228

9. Solutions of the exercises — 231

Bibliography — 261

Index — 262

1. ALGORITHMS

1.1 Algorithms and Programmes

The main activity of man is the systematic solution of problems. A problem is dealt with in two steps. First an exactly defined sequence of instructions for solving the problem is devised. This is an interesting task, which needs ingenuity. Then comes the carrying out of the instructions. Usually this is a time consuming, tedious job, which is best left to a computer. The sequence of instructions for solution of a problem is called an algorithm. The concept of algorithm is closely related to those of recipe, procedure, process, method, computer program.
Algorithms can be formulated in ordinary language. Usually we choose precise languages, which are better adapted to the representation of instruction sequences. The representation of an algorithm in a precise formal language is called a program. Construction of programs is called programming. This work aims to teach the art of programming. This art can be learned only by means of examples.
Each computer has a store which consists of locations. A location may be represented as a tiny blackboard, on which a number may be written. The computer stores the number in the form

$$a.10^b \text{ , where } b \text{ is a whole number and } 0.1 \leqslant a \leqslant 1.$$

Let the decimal representation of a be $a = 0 \cdot a_1 a_2 a_3 \ldots a_n, \; a_1 \neq 0$. Typical values for n and b are $n = 12$ and $-99 \leqslant b \leqslant 99$. If during the calculation b becomes greater than 99 or less than -99, this is described as overflow or underflow respectively. Each 'blackboard' (location) has a name - a letter, a letter followed by a number, or singly or doubly indexed letters.
Examples: A, K, S, B0, X7, P9, X(0), X(1), X(2) . . . . . . R(3, 4), etc.
The computer can carry out a limited range of simple instructions or commands.
A fundamental instruction is that of assignment to a variable. This is indicated by the symbol " ⟵ " or " := " and the instruction is frequently put in a rectangular frame. The symbol " ⟵ " or " := " is called an assignment operator.
By the instruction

A ⟵ 4

read as: 'replace A by 4', 'make A equal to 4'.

the variable A is given the value 4, i.e. the blackboard with the name A is cleaned, and the number 4 is written on it. The current value of A is 4. Suppose further that B = C = 0. Fig. 1.1 shows how the contents of A, B, C are changed by five further instructions. The so-called multiple assignment
C ⟵ B ⟵ 2A + 3 performs both the assignments B ⟵ 2A + 3 and
C ⟵ 2A + 3.

	A	B	C
$A \leftarrow A + 1$	5	0	0
$C \leftarrow B \leftarrow 2A + 3$	5	13	13
$C \leftarrow \sqrt{B^2 - A^2}$	5	13	12
$A \leftarrow (A + B + C)/(A - 2)$	10	13	12

Fig. 1.1

> Thus, the assignment
>
> Variable —— Term
>
> means:
>
> Calculate the value of the term using the current value of all variables occuring in the term. The result replaces the earlier value of the variable on the left of the arrow.

An important operation is that of exchanging the contents of two locations A and B. One might suppose that Fig.1.2 performs this. But the computer carries out the two commands consecutively. Thus the value of A, which is needed later, will be lost when the second assignment is carried out. A must first be copied on to a 'blackboard' C. Fig. 1.3 shows the correct solution. This 3-line program often occurs as part of a larger program. We shall sometimes shorten it to "A ⟷ B" (read as exchange A and B).

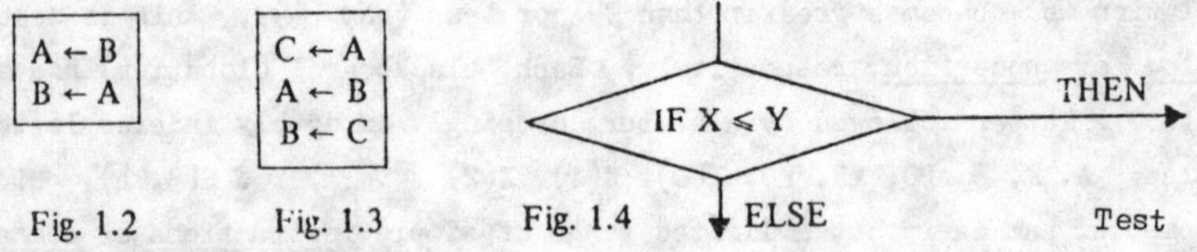

Fig. 1.2 Fig. 1.3 Fig. 1.4

Another fundamental instruction is that of decision on the basis of some comparison. This is often written in a diamond-shaped box. The truth-value (1 or 0) of one of the relations $<$, $>$, $\leqslant$, $\geqslant$, $=$, $\neq$ is tested. If the relation is true, the control of the program moves along the horizontal line to the destination given. Otherwise it passes to the next line of the instructions (Fig. 1.4). Instead of a comparison this may be referred to as a test.

The instruction PRT I (PRINT I) causes the computer to print the contents of the location I. The computer is stopped by the instruction END. Further instructions will be introduced as they become necessary. We now consider some programs.

1. Example

Fig. 1.5 shows a complete program. Lines 3 to 6 form a loop, a closed sequence of instructions. The loop contains a test which determines when the repetition is broken off. In order to find out what the program does, we run through it. The commands are carried out (executed) one after another. We start with I = 5, S = 2. At each assignment the old value is crossed out and replaced by the new one (Fig. 1.6). The program starts at I = 5 and goes in steps of S where the step length S is alternately 2 and 4. The first ten members of the sequence 6n ± 1 are printed : 5, 7, 11, 13, 17, 19, 23, 25, 29, 31.

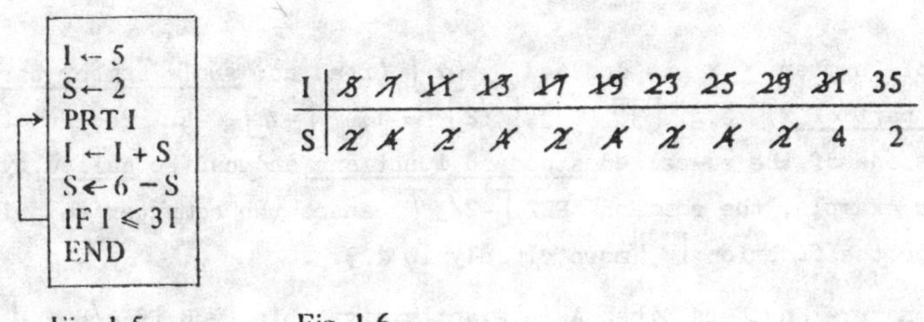

Fig. 1.5 Fig. 1.6

2. Example

Sorting three numbers. In the location A, B, C three numbers are stored. We wish to rearrange these in ascending order by pairwise exchange, and print the result. Fig. 1.7 shows a solution. An algorithm can only be understood by going through it, perhaps more than once with different data. Fig. 1.8 shows the course of the process for A = 5, B = 3, C = 1. The reader should go through the computation for these numbers and for several other orders of A, B, C.

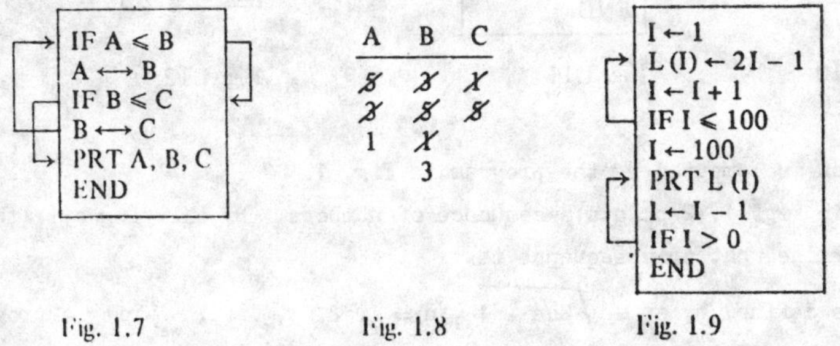

Fig. 1.7 Fig. 1.8 Fig. 1.9

3. Example

We often have to deal with lists (sequences) of numbers. Our computer permits one list for each letter of the alphabet. If L represents a list, then L(4) is the 4th element of the list. The Program in Fig. 1.9 stores the odd numbers 1, 3, 5, . . . , 199 in L(1), L(2), L(3), . . . , L(100). Then it prints these numbers in reverse order.

4. Example

Fig. 1.10 shows a program. INP A is an abbreviation for INPUT A. When this program is typed into the computer and run, the computer ouputs a question mark

?

Then we must input the value of A, for example 2. The computer then calculates a number P, determined by the value of A(=2), and prints it out. Then we can input another value for A and the corresponding value of P is calculated. That is, the program represents a function f, which calculates a value P = f(A) for each input value of A.

5. Example

The largest whole number $\leq x$ is denoted by $[x]$ (read as: whole number part of x, or integer part of x) e.g. $[3] = 3$, $[\sqrt{2}] = 1$, $[-\pi] = -4$. The function integer part is one of the so-called standard functions and can be called by using its name. For example, the command PRT $[-2/3]$ causes the computer to print -1. We shall examine the function $[\;]$ more closely in 1.3.3.

If A and B are natural numbers, then A is exactly divisible by B if $A/B = [A/B]$. The program in Fig. 1.11 prints all odd numbers from 11 to 119 which are divisible neither by 3, nor by 5, nor by 7, that is, precisely the prime numbers in this interval.

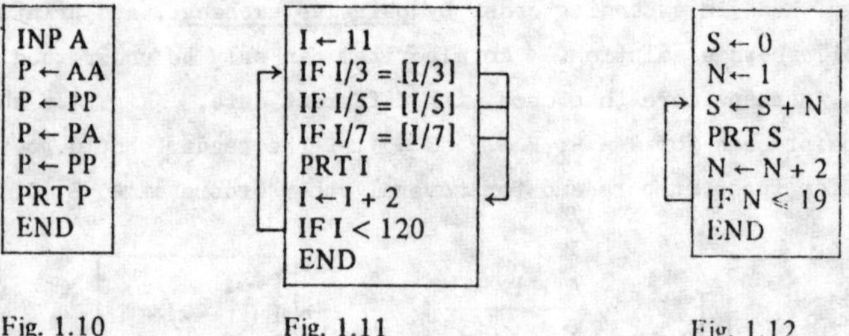

Fig. 1.10 Fig. 1.11 Fig. 1.12

Exercises

1. Which function is computed by the program in Fig. 1.10?
2. The program in Fig. 1.12 prints a sequence of numbers. By carrying-out the program determine what this sequence is.
3. A sequence is defined by $a_n = \sqrt{24n + 1}$, n = 1, 2, 3, . . . Write a program which prints the first ten whole-number members of this sequence with their position in the sequence. Run the program and examine the values printed out. Make a conjecture!

   Hint. a_n is a whole number if and only if $a_n = [a_n]$.
4. Construct an algorithm, by analogy with Fig. 1.7, which sorts four numbers into order.

5. It is required to transform the sequence (A, B, C, D, E) into the sequence (B, C, D, E, A) using the least possible number of instructions (cyclic interchange).

6. What is the outcome of the three instructions

   $A \leftarrow A + B$; $B \leftarrow A - B$; $A \leftarrow A - B$?

7. The program in Fig. 1.13 prints a sequence of numbers. By carrying out the program identify the sequence.

8. <u>The Egyptian method of multiplication.</u> The ancient Egyptians knew well how to add, double and halve. From this they could form the product Z of two numbers X and Y. They used the identity $XY = X(Y - 1) + X = (2X)(Y/2)$ in order to reduce Y step-by-step, as shown by the following numerical example:

   25.19 = 25.18 + 25 = 50.9 + 25 = 50.8 + 75 = 100.4 + 75 = 200.2 + 75 = 400.1 + 75 = 475.

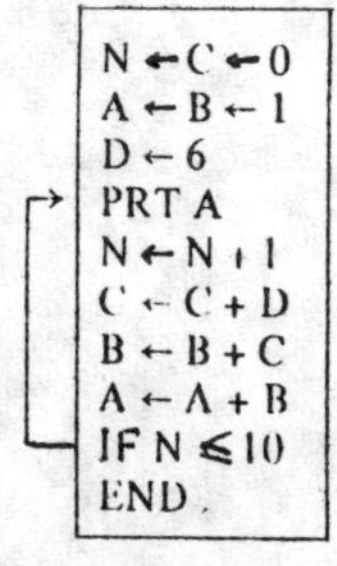

Fig. 1.13

X ← 2X	Y ← [Y/2]
25	19
50	9
~~100~~	4
~~200~~	2
400	1
475	

Fig. 1.14 a

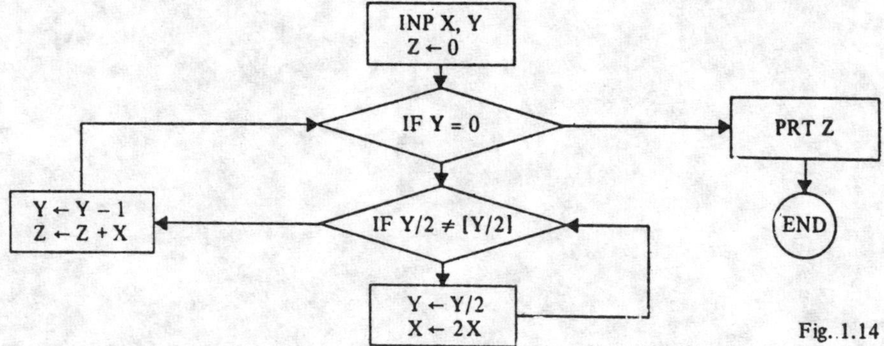

Fig. 1.14

a) Study this example and compare it with the corresponding program in Fig.1.14.

b)] Simplify the multiplication procedure in the manner of the table 1.14a. Simplify the program in Fig. 1.14 in the same way.

[9.] Fig. 1.15 shows an interesting program. A real number X is input. The output is the number Y = f(X). A and B are auxiliary variables. Determine the function f.

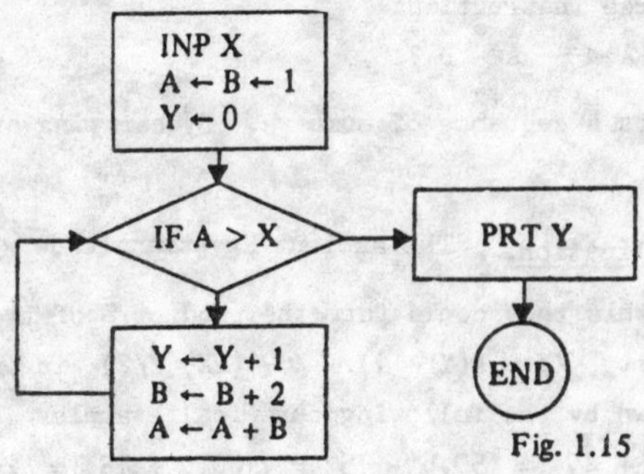

Fig. 1.15

1.2 The 3A + 1 Problem. Empirical investigation of an algorithm

A few years ago there appeared an interesting algorithm for producing a number sequence. It goes as follows:

1. Start with an arbitrary natural number A.
2. If A = 1, then stop.
3. If A is even, replace A by A/2 and go to step 2.
4. If A is odd, replace A by 3A + 1 and go to step 2.

We carry out the algorithms for starting values 3, 34, 75:

3, 10, 5, 16, 8, 4, 2, 1.

34, 17, 52, 26, 13, 40, 20, 10, 5, 16, 8, 4, 2, 1

75, 226, 113, 340, 170, 85, 256, 128, 64, 32, 16, 8, 4, 2, 1

Does the algorithm stop after a finite number of steps for every starting value? We shall collect some empirical data. The computer will be instructed to print the corresponding sequence for any input value A.

Until now we have written all programs one-dimensionally, that is, the rows of instructions follow each on a new line. A so-called <u>flow diagram</u> (Fig. 1.16) is somewhat clearer, since it is two dimensional. Our programs, whether one- or two-dimensional, are independent of a particular language or machine. To translate them into any desired computer language is a trivial exercise. We shall also use the computer language BASIC since it is widely used in schools and available on micro-computers.

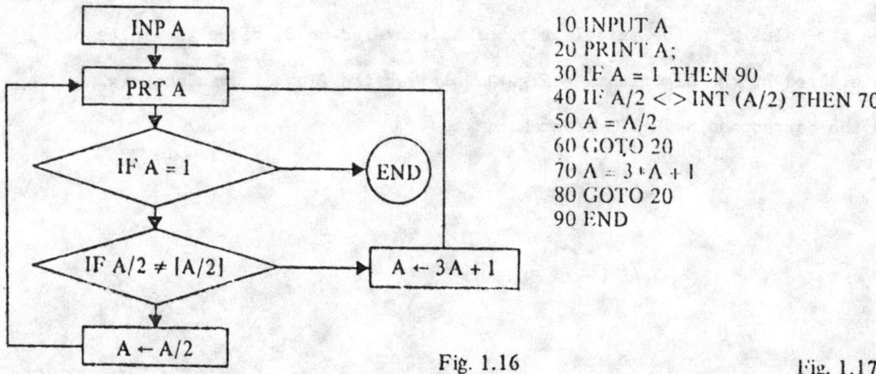

Fig. 1.16 Fig. 1.17

Fig. 1.17 shows the BASIC program. The BASIC language is almost self-explanatory. It is most easily learnt by examining a few programs. Fig. 1.17 shows that the function "[ ]" is written as "INT" in BASIC (integer part). Instead of A ⟵ B we use A = B or LET A = B. The multiplication sign * must not be omitted. Line 20 needs some explanation. Of the three commands :-

20 PRINT A   20 PRINT A,   20 PRINT A;

the first prints one member of the sequence on each line, the second five members per line and the third as many as there is room for. Instead of " ≠ " the symbol "<>" is used. The lines must be numbered in ascending order. It is convenient to use the numbers 10, 20, 30, . . . If later one wishes to insert another instruction between line 30 and 40, it may be given any of the numbers between 30 and 40 and typed in at the end.

The input A = 27 gives the sequence:

27 82 41 124 62 31 94 47 142 71 214 107 322 161 484 242 121 364
182 91 274 137 412 206 103 310 155 466 233 700 350 175 526 263 790
395 1186 593 1780 890 445 1336 668 334 167 502 251 754 377 1132
566 283 850 425 1276 638 319 958 479 1438 719 2158 1079 3238 1619
4858 2429 7288 3644 1822 911 2734 1367 4102 2051 6154 3077 9232 4616
2308 1154 577 1732 866 433 1300 650 325 976 488 244 122 61 184 92
46 23 70 35 106 53 160 80 40 20 10 5 16 8 4 2 1

This is an obscure deluge of numbers. It is more instructive to print not the whole sequence but only the starting value and the number of steps to reach the value 1. The inital value of A is stored as B, since the value of A is altered by the algorithm. Fig. 1.18 gives the flow diagram and Fig. 1.19 the corresponding BASIC program.

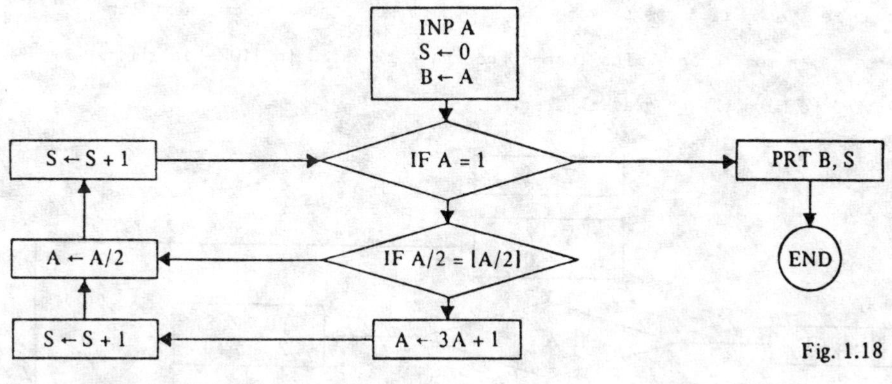

Fig. 1.18

```
 10 INPUT A
 20 S = 0
 30 B = A
 40 IF A = 1 THEN 110
 50 IF A/2 = INT (A/2) THEN 80
 60 A = 3*A + 1
 70 S = S + 1
 80 A = A/2
 90 S = S + 1
100 GOTO 40
110 PRINT B; S
120 END
```

Fig. 1.19

We use the computer to print a table which gives the corresponding number of steps S for each input value A from 1-50. Perhaps we shall then observe some regularity.

Fig. 1.20 arises from an easily understood extension of Fig.1.18. The variable Z is a counter, which runs from 1 to 50. The corresponding lines

```
10 FOR Z = 1 to 50 ......
120 NEXT Z
```

are a repeat instruction or a FOR-loop. Their effect is to cause that part of the program lying between them to be repeated 50 times. At each run through, the counter Z is increased by 1. When the computer reads line 10 it sets Z = 1 and notes the upper limit 50. Then it carries out lines 20 to 110. At 120 it increases Z by 1 and tests whether $Z \leq 50$. If the answer is 'yes', it jumps back to 20. Otherwise it goes to the next line and so comes to a stop. The loop structure - lines 20 to 110 - is shown indented to make the program clearer.

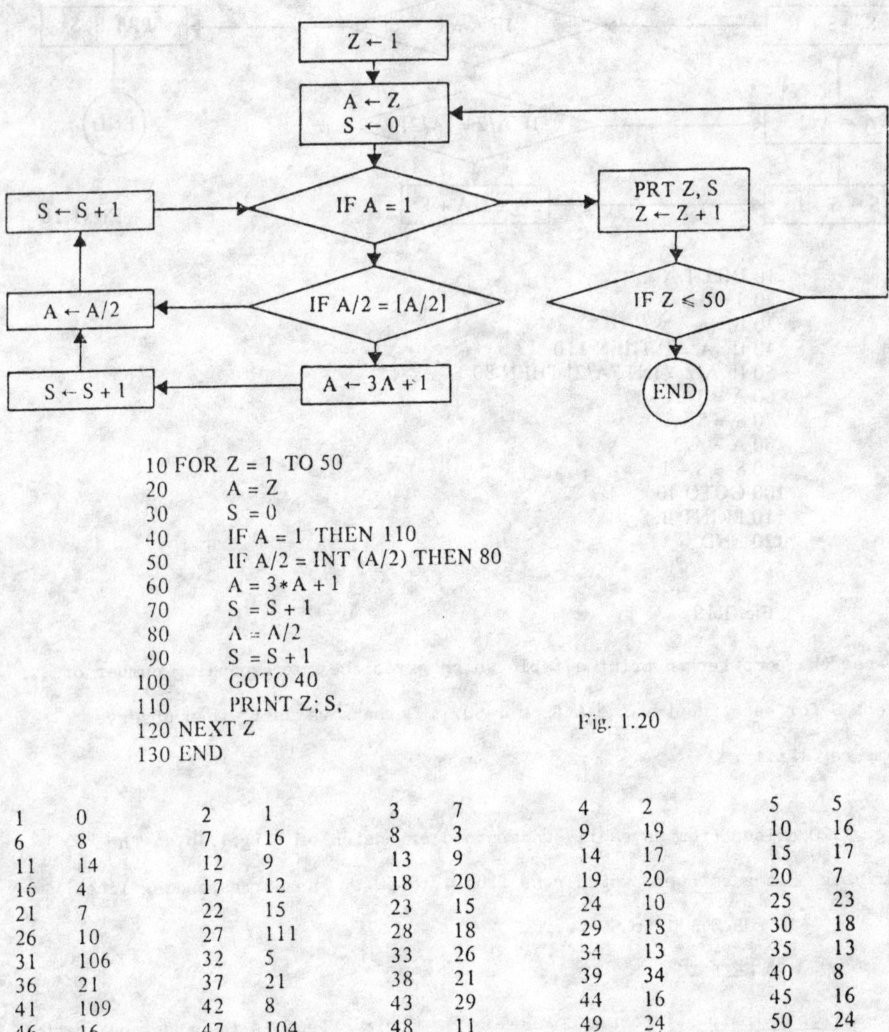


```
10 FOR Z = 1 TO 50
20       A = Z
30       S = 0
40       IF A = 1  THEN 110
50       IF A/2 = INT (A/2) THEN 80
60       A = 3*A + 1
70       S = S + 1
80       A = A/2
90       S = S + 1
100      GOTO 40
110      PRINT Z; S,
120 NEXT Z
130 END
```

Fig. 1.20

1	0	2	1	3	7	4	2	5	5
6	8	7	16	8	3	9	19	10	16
11	14	12	9	13	9	14	17	15	17
16	4	17	12	18	20	19	20	20	7
21	7	22	15	23	15	24	10	25	23
26	10	27	111	28	18	29	18	30	18
31	106	32	5	33	26	34	13	35	13
36	21	37	21	38	21	39	34	40	8
41	109	42	8	43	29	44	16	45	16
46	16	47	104	48	11	49	24	50	24

The program in Fig. 1.20 is not fool-proof. It is possible that the number A might become so large during the computation that it would be rounded off. The incorrect value would then be used in further computation and incorrect results printed. We can guard against this danger by also printing the largest member M of the sequence which is generated. Then we can see whether M falls into the danger zone near 10^{12}. The variable M in Fig. 1.21 represents the current maximum. Initially M is set at 0. Then every member of the sequence is compared with M, and if $A > M$, M is replaced by A ($M \leftarrow A$). After the step $A \leftarrow A/2$ a comparison of M and A is unnecessary. In Fig. 1.21 we use for the first time a connector (1) which indicates where the flow diagram continues. In this way one can avoid complicated cross overs. This time the counter Z runs from X to Y. The bounds X and Y must be given. In Fig. 1.21 they are chosen as X = 71, Y = 100.

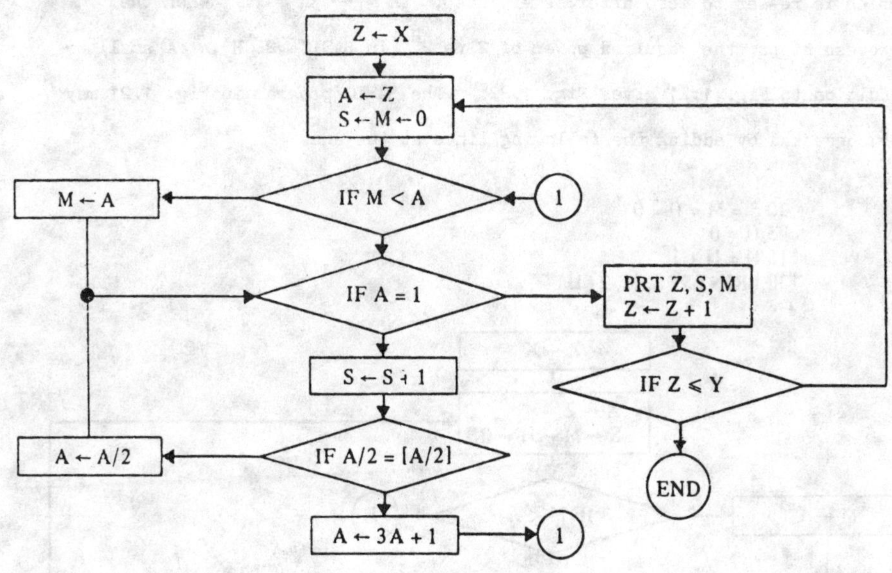


```
10 FOR Z = 71 TO 100
20        A = Z
30        S = M = 0
40        IF M > = A THEN 60
50        M = A
60        IF A = 1 THEN 130
70        S = S + 1
80        IF A/2 = INT (A/2) THEN 110
90        A = 3*A + 1
100       GOTO 40
110       A = A/2
120       GOTO 60
130       PRINT Z; S; M,
140 NEXT Z
150 END
```

Fig. 1.21

71	102	9232	72	22	72	73	115	9232	74	22	112	75	14	340
76	22	88	77	22	232	78	35	304	79	35	808	80	9	80
81	22	244	82	110	9232	83	110	9232	84	9	84	85	9	256
86	30	196	87	30	592	88	17	88	89	30	304	90	17	136
91	92	9232	92	17	160	93	17	280	94	105	9232	95	105	9232
96	12	96	97	118	9232	98	25	148	99	25	448	100	25	100

Fig. 1.21

We wish to improve this program further. The algorithm only comes to a halt when a power of 2 is produced. From then on the sequence decreases by successive halving until 1 is reached. We introduce the counter H (initially H = 0) which is increased by 1 after every step $A \leftarrow A/2$, and which is re-set to zero after every step. $A \longleftarrow 3A + 1$. When the process stops, the required power of 2 is 2^H (in BASIC 2↑H). A small addition to Fig. 1.21 gives Fig. 1.22. The BASIC program in Fig. 1.21 may be corrected by adding the following lines at the end.

```
30 S = M = H = 0
95 H = 0
115 H = H + 1
130 PRINT Z, S, M, 2↑H
```

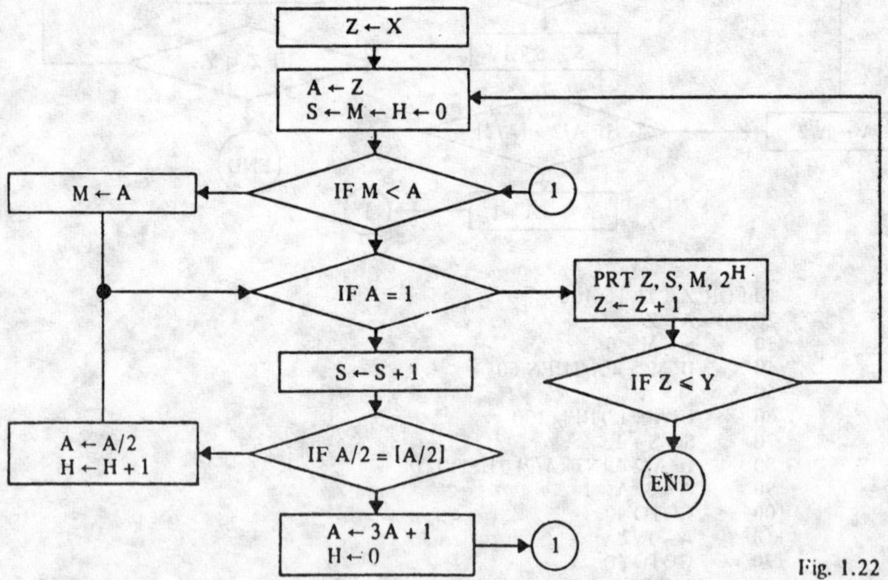

Fig. 1.22

It is not known whether for arbitrary initial values the sequence diverges to ∞ or goes into a loop. The Artificial Intelligence Laboratory at M.I.T. (Memo 239, 1972) has investigated all A for $-10^{-8} < A \leq 6 . 10^{7}$. Even values of A were replaced by A/2 and odd values by 3A + 1, iteratively, (as above). The sequence always ended in one of the following five loops.

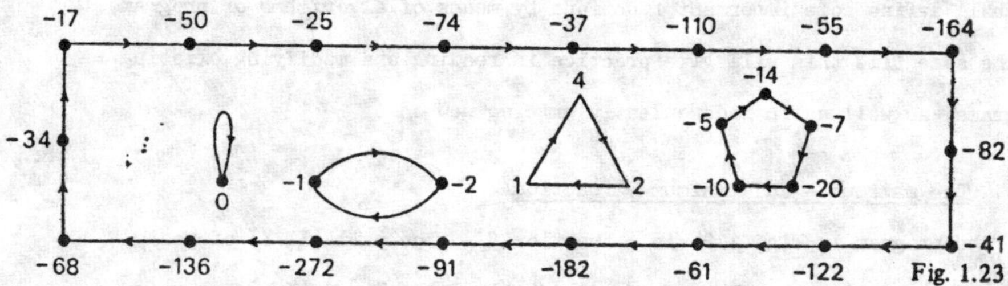

Fig. 1.23

D. J. Selfridge (Berkeley) has verified that the algorithm stops for all $A \leq 2^{29}$. According to an unconfirmed report this is true for all $A \leq 10^{40}$

Exercises

1. Try the program in Fig. 1.22 for X = Y = 31466 383 (Result: S = 705).
2. Simplify Fig. 1.22 so that for each value of A the program determines only the first power of 2, 2^H, encountered, and prints (A, 2^H).
3. In the 3A + 1 problem, for each initial value A let S(A) be the number of steps before the program stops. A is to run through the odd numbers from 1 to 1161. The pair (A, S(A)) is to be printed only when S(A) is a record, i.e. larger than any preceding value of S(A). Write a program to do this (by modifying Fig. 1.20) and run it on the computer.
4. Here is an algorithm which produces a number sequence;
   1. Start with an arbitrary natural number A.
   2. If A = 4, then STOP.
   3. If A ends in the digit 4, cross out the 4 and go to step 2.
   4. If A ends in the digit 0, cross out the 0 and go to step 2.
   5. Double A and go to step 2.

   Draw a flowchart, translate in into BASIC and experiment with several initial values. The value 1249 is particularly interesting.

Hint. B = A - 10 [A/10] is the remainder when A is divided by 10 i.e. the last digit of A. In BASIC, the function |X| is written ABS (X). Using this, the program can be simplified somewhat.

1.3 Algorithmic definition of a function

We shall define some important functions by means of algorithms or programs. At the same time this will give practice in reading and modifying existing programs, as well as in independently writing new ones.

1.3.1 The maximum - and minimum - function

a) The program in Fig. 1.24 is a function f. The pair (X, Y) of real numbers is input. The program then calculates Z = f (X,Y).

e.g. f(3,5) = f(5,3) = 5 f(3,3) = 3. Clearly Z = max (X,Y).

b) The function g in Fig.1.25 calculates from the number triple (X,Y,Z) a number M.

e.g. g(3,4,5) = g(5,3,4) = 5. Here M = max (X,Y,Z).

c) Numbers are stored in the locations R1, R2,. . ., R(N). In which location I is the largest M, and how large is M?

We must make the question more precise. The largest number could occur in several locations. We seek that largest number which lies furthest to the left. i.e. we seek

$$M = \max_{1 \le J \le N} R(J) \quad \text{and} \quad I = \min \left\{ J \,/\, R(J) = M \right\}$$

The figures 1.26 and 1.27 show the maximum-program and a development. I is the number of the location which holds the current maximum (initially, I = 1). K is the number of the location whose content must be compared with the current maximum. (initially, K = 2).

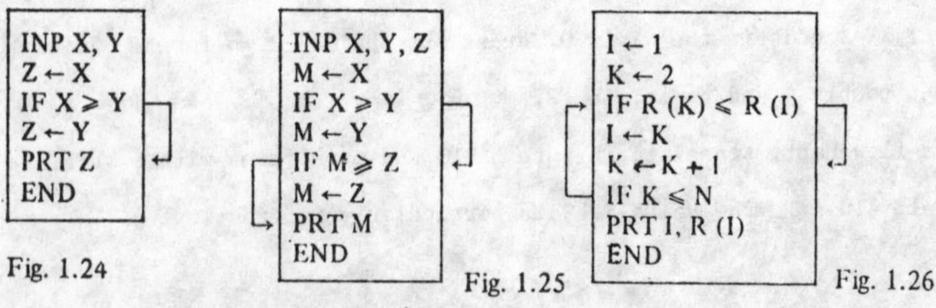

Fig. 1.24 Fig. 1.25 Fig. 1.26

R (1)	R (2)	R (3)	R (4)	R (5)	R (6)
2	4	6	3	6	3

N	6
I	~~1~~ ~~2~~ 3
K	~~2~~ ~~3~~ ~~4~~ ~~5~~ ~~6~~ 7

Fig. 1.27

Exercises

1. Show that $\max(x,y) = \frac{x+y+|x-y|}{2}$, $\min(x,y) = \frac{x+y-|x-y|}{2}$

2. The sum function is one of the standard functions of the computer, i.e. the assignment $m \leftarrow \frac{x+y+|x-y|}{2}$ gives max (x,y). How can max (x,y,z) be calculated using two assignments, and max(x,y,z,u) using three?

3. Alter the programs in Figs. 1.24 to 1.26 so that they determine minimum values instead of maximum values.

4. Suppose the addition function was not available as a standard computer function. How would you write a program which outputs $|x|$ when x is input?

5. Alter Fig. 1.26 so that

   a) the largest number which lies furthest to the <u>right</u> is printed.

   [b)] all pairs (I,M) with maximal M are printed.

6. Translate Fig. 1.26 into BASIC.

7. Write a BASIC program which determines the maximum and the minimum values in the sequence R(1), R(2), . . . R(N).

8. Write a BASIC program which determines the maximum (minimum) value in the sequence $R(I) = I\sqrt{2} - [I\sqrt{2}]$, I = 1, 2, . . . 100.

1.3.2. Factorials and Powers

The function $n \rightarrow n!$ is defined by $0! = 1! = 1$, $n! = 1.2.3.4\ldots n$.

The symbol n! is read as "n factorial". The program in Fig. 1.28 gives the output (N,N!) when N is input.

Exponentiation (raising to a power), X^Y, where $X > 0$, is one of the operations which the computer can perform. We will write our own exponentiation program, since we will learn a lot by doing so. Fig. 1.29 produces the output $P = A^N$, when A and N are input, provided $N \in \{0,1,2,\ldots\}$.

The program may be tested for

a) A = 2, N = 6 b) A = 2, N = 1 c) A = 2, N = 0

The program is wasteful; for example it uses 16 multiplications to calculate A^{16}. We can do this in four multiplications, obtaining A^{16} by squaring four times.

$P \leftarrow AA$; $P \leftarrow PP$; $P \leftarrow PP$; $P \leftarrow PP$

The problem of optimal calculation of an exponent is difficult. A M Legendre (Theorie des nombres, 1798) found an elegant algorithm, which is almost optimal. The basic idea is illustrated by the following example:

$1.2^{15} = 2.2^{14} = 2.4^{7} = 8.4^{6} = 8.16^{3} = 128.16^{2} = 128.256 = 32768$

Here two steps are used each of which leaves the product $Z.X^Y$ unchanged.

Step 1: $Y \leftarrow Y - 1$, $Z \leftarrow Z.X$

Step 2: $Y \leftarrow Y/2$, $X \leftarrow X^2$

The second step is only possible when Y is even. If Y is odd, then it is made even by the first step. We start with X = A, Y = N, Z = 1. Then the product $Z.X^Y$ is constant and equal to A^N. When Y is even the second step is used and when Y is odd the first step is used. The exponent Y always decreases, and when Y = 0, $Z = A^N$. Fig. 1.30 shows the corresponding program. Compare Fig. 1.30 with Fig. 1.14.

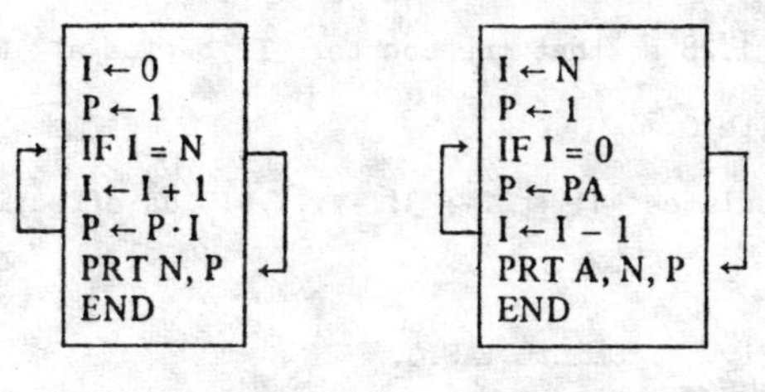

Fig. 1.28 Fig. 1.29

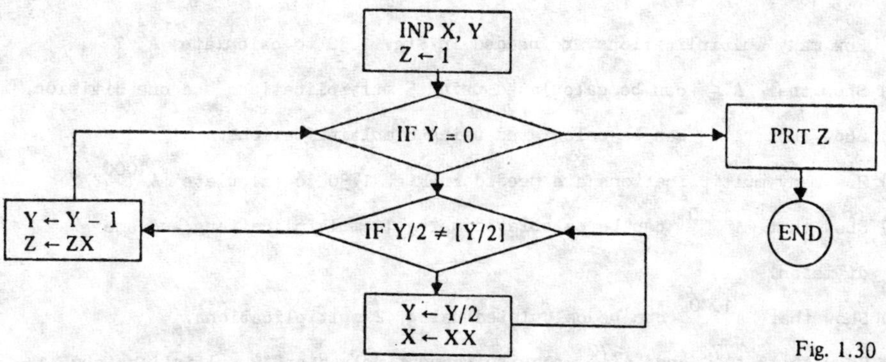

Fig. 1.30

The program uses 7 multiplications to calculate A^{15} (or 6 if multiplication by 1 is discounted). The optimal program uses only 5 multiplications in calculating A^{15}, as Fig. 1.31 shows. The program in Fig. 1.32 squares four times and finally divides by A, but division is more expensive than multiplication.

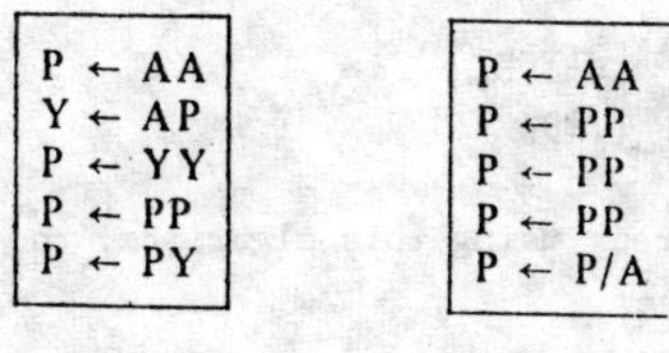

Fig. 1.31 Fig. 1.32

Exercises

1. Try out the program in Fig. 1.28 for N = 6, 4, 1, 0.
2. Alter the program in Fig. 1.28 so that a table of values of the function I → I!, for I = 0, 1, 2,... N, is printed.

3. Alter the program in Fig. 1.28 so that the counter I begins at N and counts down, ending with 0.

4. Write a program which calculates 1! + 2! + 3! + ... N! as effectively as possible.

5. Translate the program in Fig. 1.30 into BASIC.

6. Use the program in Fig. 1.30 to calculate $(1 - \frac{1}{n})^n$ for n = 10, 10^2 .. ... 10^{12}.

7.a) How many multiplications are needed in Fig. 1.30 to calculate A^{23}?

b) Show that A^{23} can be calculated using 5 multiplications and one division.

c) Show that A^{23} can be calculated using 6 multiplications.

8.a) How many multiplications are needed in Fig. 1.30 to calculate A^{1000}?

b) Show that A^{1000} can be calculated using 11 multiplications and one division.

c) Show that A^{1000} can be calculated using 12 multiplications.

9. Show that A^{77} and A^{170} require respectively 8 and 9 multiplications.

10. Here is an algorithm for calculating A^N. Write N in binary notation. Replace each "1" by "SA" and each "0" by "S". Delete SA from the left-hand end. The list remaining is a set of instructions for calculating A^N, if S and A are interpreted as the commands "square" and "multiply by A" respectively.

Example: $23 = 10111_2$ = SASSASASA.

The series of instructions SSASASA gives

$A \to A^2 \to A^4 \to A^5 \to A^{10} \to A^{11} \to A^{22} \to A^{23}$

How many multiplications are required, using this algorithm, to calculate A^{15}, A^{16} and A^{1000} ?

1.3.3. The function "integer part"

We now examine one of the most important standard functions of the computer. It is used in most programs. We define

$[x]$ = the greatest integer $\leq x$, $x \in \mathbb{R}$.

The symbol "$[x]$" is read as "the integer part of x". In BASIC this function is called INT. Fig. 1.33 shows the graph of the function $x \to [x]$.

Remarks and Examples

a) $[x]$ arises from rounding-down x to an integer. For example:

$[3.7] = 3$, $[4] = 4$, $[-3.7] = -4$, $[-5] = -5$, $[0.6] = 0$, $[-0.6] = -1$

b) $[X] = X$ is true if and only if X is an integer.

Thus the conditional

IF X = INT (X)

is a test to determine whether X is integral.

c) Suppose N and P are integers. Then $[N/P] = N/P$ if and only if P is a divisor of N. i.e.

IF N/P = INT (N/P)

is a test for divisibility of N by P.

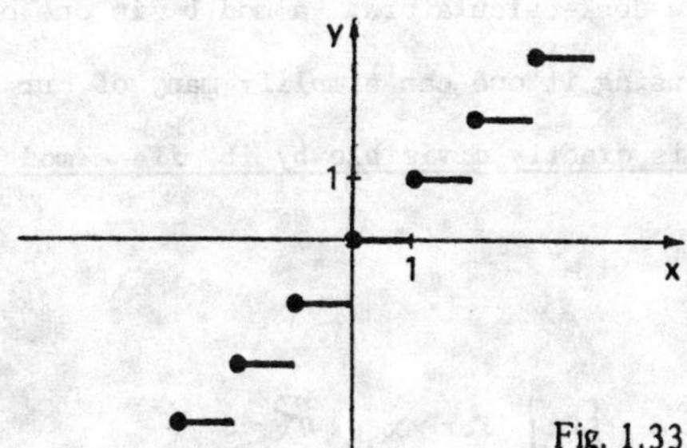

Fig. 1.33

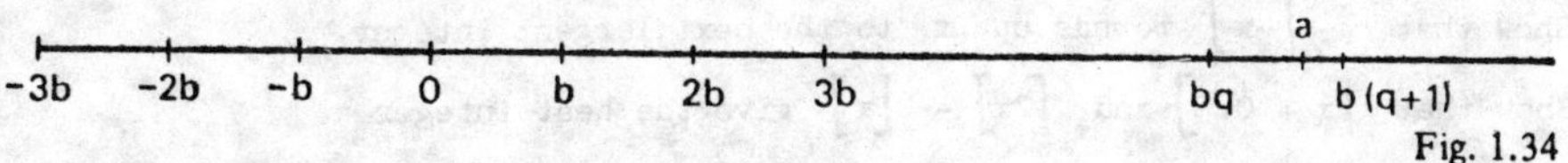

Fig. 1.34

d) Division with remainder

An integer a is uniquely expressible in the form

(1) $a = bq + r$, $0 \leqslant r < b$, where b is a natural number. q and r are called the quotient and remainder in the division of a by b. In order to see this we consider the whole number multiples of b on the number-line (Fig. 1.34). The integer a falls in exactly one of the intervals $[bq, b(q+1)]$. From (1) it follows that

$$\frac{a}{b} = q + \frac{r}{b}, \quad 0 \leq \frac{r}{b} < 1$$

Therefore $q = \left[\frac{a}{b}\right]$, $r = a - b\left[\frac{a}{b}\right]$

The expression

$a - b\left[\frac{a}{b}\right] =$ remainder when a is divided by b

is often useful and should be remembered.

It is convenient to define for all real numbers a, b the expression $a \bmod b$ (read as <u>a modulo b</u>) as follows

$$a \bmod b = \begin{cases} a - b\,[a/b] & \text{for } b \neq 0 \\ 0 & \text{if } b = 0 \end{cases}$$

In particular

$a \bmod 1 = a - [a]$ = fractional part of a.

In large computers and in a few desk-calculators, $a \bmod b$ is one of the standard operations. By using it one can simplify many of our programs. In particular, <u>a is exactly divisible by b if a mod b is zero</u>.

<u>Exercises</u>

1. Determine all solutions of $2x = [2x]$ for $x \in \mathbb{R}$.
2. For which n is the condition $\sqrt{n} = [\sqrt{n}]$ true?
3. Show that $-[-x]$ rounds up x to the next largest integer.
4. Show that $[x + 0.5]$ and $[2x] - [x]$ give the best integer approximation to x, i.e. they round x to the nearest integer. (Rounding function).
5. Sketch the graphs of the following functions.

   a) $y = \frac{[2x]}{2}$ b) $y = 2\left[\frac{x}{2}\right]$ c) $y = (-1)^{[x]}$ d) $y = (-1)^{[2x]}$

   e) $y = x - [x]$ f) $y = -[-x]$
6. Calculate $\frac{[10x + 0.5]}{10}$, $\frac{[100x + 0.5]}{100}$, $\frac{[1000x + 0.5]}{1000}$ for $x = 3.1416$ and $x = 2.71828$. Hence interpret $\frac{[10^d x + 0.5]}{10^d}$ for $d = 0, 1, 2, 3, \ldots$

7. The following formula, due to Rev. Zeller, gives the day of the week for each date of the Gregorian Calendar

$$W = D + [2.6M - 0.2] + Y + [Y/4] + [C/4] - 2C$$

where C is the century, Y the year number within the century, M the month, D the day, and $W \bmod 7 = W - 7[W/7]$ is the day of the week (where Sunday = 0, Monday = 1. . . Saturday = 6). The Roman numbering of months is used i.e. March is M = 1, April is M = 2...., January and February are the 11th and 12th months of the year.

Calculate the day of the week for the following dates:

a) today b) your birthday c) 1.1.2000 d) 22.6.1941 e) 7.12.1941 f) 25.6.1950

Example:

For 1.2.1902 we have C = 19, Y = 2, M = 12, D = 1, W = -1, W mod 7 = 6.

8.a) If N runs from 1 to A^2, then $f(N) = [N + \sqrt{N} + \frac{1}{2}]$ runs to $A^2 + A$, omitting precisely A numbers. Write a program which prints out the numbers omitted. Make a conjecture.

b) Let N run through the numbers 1, 2, 3, 4. . . Which numbers do not appear as values of the function $g(N) = [N + \sqrt{2N} + \frac{1}{2}]$?

9. Let $x = z_n z_{n-1} \ldots z_2 z_1 z_0, z_{-1} z_{-2} \ldots$ be the decimal representation of a real number $x \geqslant 0$. Then $z = \left[\frac{x}{10^i}\right] - 10\left[\frac{x}{10^{i+1}}\right]$ gives the ith digit of x. Prove this.

10. In Fig. 1.35 a natural number N is input. How does the output M depend on N?

```
   INP N
   M ← Z ← 0
→  Z ← N − 10 [N/10]
|  N ← [N/10]
|  M ← 10M + Z
└─ IF N > 0
   PRT M
   END
```

Fig. 1.35

1.34 Square root and other functions

a) The square root algorithm. Over 4000 years ago the Sumerians knew a method of calculation which given an input $a > 0$ would produce the output $\sqrt{a}$. It will be mentioned only briefly here since it is dealt with more fully in Chapter 4.

Any value $x_0 > 0$ is chosen, and $x_1, x_2 \ldots$ are found from the recursive formula

$$(1) \quad x_{n+1} = \frac{1}{2}\left(x_n + \frac{a}{x_n}\right)$$

We show that $x_1 > x_2 > x_3 \quad \ldots\ldots > \sqrt{a}$ and $\lim_{n\to\infty} x_n = \sqrt{a}$.

Writing

$$(2) \quad x_n = \sqrt{a}(1 + \varepsilon_n),$$

where $\varepsilon_n = \frac{(x_n - \sqrt{a})}{\sqrt{a}}$ is the nth relative error, it follows from (1) and (2) after a little calculation that

$$x_{n+1} = \sqrt{a}\left(1 + \frac{\varepsilon_n^2}{2(1 + \varepsilon_n)}\right)$$

i.e.

$$(3) \quad \varepsilon_{n+1} = \frac{\varepsilon_n^2}{2(1 + \varepsilon_n)}$$

Since $x_0 > 0$, we have $\varepsilon_0 > -1$ and so $\varepsilon_n > 0$ for all $n \geqslant 1$.

i.e. $x_n > \sqrt{a}$ for $n \geqslant 1$. By ignoring in turn 1 and ε_n in (3), one obtains, for $n > 1$,

$$(4) \quad \varepsilon_{n+1} < \frac{\varepsilon_n}{2}$$

$$(5) \quad \varepsilon_{n+1} < \frac{\varepsilon_n^2}{2}$$

By (4) the relative error is more than halved at each step. If $\varepsilon_n < 10^{-p}$ then by (5) $\varepsilon_{n+1} < \frac{10^{-2p}}{2}$ i.e. at each step the number of correct decimal digits is approximately doubled.

Fig. 1.36 shows the program for calculating $\sqrt{a}$. We have chosen $x_0 = \frac{1 + a}{2}$; y and x are two successive members of the sequence x_n. The sequence $x_1, x_2, x_3, \ldots$ is printed so long as $x_{n+1} < x_n$.

Initially this condition is satisfied, but in the neighbourhood of $\sqrt{a}$ it is not fulfilled because of round-off errors. Then it is time to stop. The input $a = 2$ gives $\sqrt{2} = 1.414213562$ after 3 steps, which is correct to 10 digits. (Fig. 1.36a)

```
X ← (1 + A)/2
Y ← X
X ← (Y + A/Y)/2
PRT X
IF X < Y
END
```

Fig. 1.36

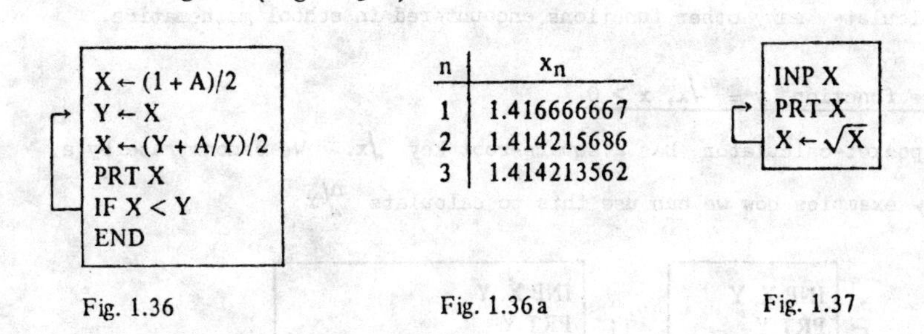

n	x_n
1	1.416666667
2	1.414215686
3	1.414213562

Fig. 1.36 a

```
INP X
PRT X
X ← √X
```

Fig. 1.37

b) <u>Continued extraction of square roots</u>

Fig. 1.37 shows a small program. If a number $x > 0$ is input, it prints the sequence $x_o = x$, $x_1 = \sqrt{x_o}$, $x_2 = \sqrt{x_1}$, . . . $x_{n+1} = \sqrt{x_n}$, . . .

Thus $x_n = x_{n+1}^2$

We show that $\lim_{n \to \infty} x_n = 1$. Suppose first that $x > 1$. Then $x_n > 1$ for all n, and

$$(6) \quad \frac{x_{n+1} - 1}{x_n - 1} = \frac{x_{n+1} - 1}{x_{n+1}^2 - 1} = \frac{1}{1 + x_{n+1}} < \frac{1}{2}$$

The transition from x_n to x_{n+1} is called one step or one iteration. (3) shows that each step more than halves the distance from 1. It follows that $x_n \to 1$ as $x \to \infty$. If one starts with a large number e.g. $x = 10^{10}$ the number of digits is approximately halved at each step, and one soon obtains $x_n < 10$. After at most two more iterations we have $1 < x_n < 2$. From that point the convergence is slow. At each iteration, the distance from 1 is more nearly halved, for

$$\lim_{n \to \infty} \frac{x_{n+1} - 1}{x_n - 1} = \lim_{n \to \infty} \frac{1}{1 + x_{n+1}} = \frac{1}{2}$$

Thus we say x_n converges <u>linearly</u> to 1 with <u>convergence factor</u> $\frac{1}{2}$.

Finally, if $0 < x < 1$, then $y = \frac{1}{x} > 1$ and the sequence $y_n = \frac{1}{x_n}$ converges to 1, and therefore x_n does so too.

If one starts in the interval $\frac{1}{2} < x < 2$, after 32 iterations we have $|1 - x_{32}| \leqslant 1.614 \cdot 10^{-10}$. This can be checked using a hand-calculator. See also Exercise 4. Using continued square-root extraction we can calculate many other functions encountered in school mathematics.

c) <u>The function $y = \sqrt[n]{x}$, $x > 0$</u>

A pocket-calculator has a square-root key $\sqrt{x}$. We demonstrate by a few examples how we can use this to calculate $\sqrt[n]{x}$

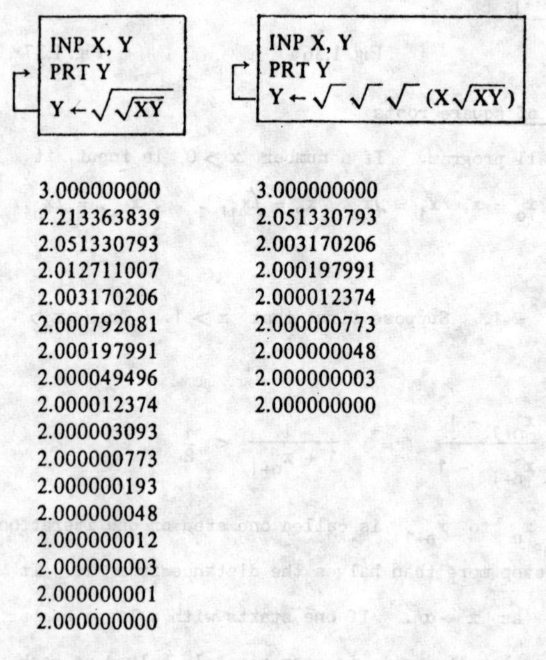

Fig. 1.38	Fig. 1.39
3.000000000	3.000000000
2.213363839	2.051330793
2.051330793	2.003170206
2.012711007	2.000197991
2.003170206	2.000012374
2.000792081	2.000000773
2.000197991	2.000000048
2.000049496	2.000000003
2.000012374	2.000000000
2.000003093	
2.000000773	
2.000000193	
2.000000048	
2.000000012	
2.000000003	
2.000000001	
2.000000000	

Fig. 1.38 Fig. 1.39

1. <u>Example</u>

Let x be given with $x \geqslant 0$. To find $\sqrt[3]{x}$. For $y > 0$ we have

$y = \sqrt[3]{x} \Leftrightarrow y^3 = x \Leftrightarrow y^4 = xy \Leftrightarrow y = \sqrt{\sqrt{xy}}$.

We seek the solution of the equation $y = \sqrt{\sqrt{xy}}$.

If in Fig. 1.38 the value of x and an estimate y_0 for the unknown y are input, the output is the sequence

$y_0,\ y_1 = \sqrt{\sqrt{xy}},\ y_2 = \sqrt{\sqrt{xy_1}}\ \ldots\ y_{n+1} = \sqrt{\sqrt{xy_n}}\ \ldots$

We assert that $y_n \to y$ as $n \to \infty$ with convergence-factor $\frac{1}{4}$. In fact, if the approximation y_0 has an error factor q_0, i.e. $y_0 = yq_0$, then we have $y_1 = \sqrt{\sqrt{xyq_0}} = \sqrt{\sqrt{xy}}\sqrt{\sqrt{q_0}} = yq_1$

i.e. y_1 has the error-factor $q_1 = \sqrt{\sqrt{q_0}}$.

Similarly, $y_2 = yq_2$, $y_3 = yq_3$, etc. with $q_{n+1} = \sqrt{\sqrt{q_n}}$

From b) q converges to 1 with convergence-factor $\frac{1}{4}$, since in the transition from q_n to q_{n+1} the square-root is extracted twice.

Fig. 1.38 shows the calculation of $\sqrt[3]{8}$, starting from the estimate $y_0 = 3$. At each iteration the distance from the limiting value $y = 2$ is more and more exactly four times smaller. After 16 iterations the error is less than 10^{-10}.

2. Example

We wish to calculate $y = \sqrt[5]{x}$ for $x > 0$. For $y > 0$,

$$y = \sqrt[5]{x} \iff y^5 = x \iff y^{15} = x^3 \iff y^{16} = x^3 y \iff y = \sqrt{\sqrt{\sqrt{x\sqrt{xy}}}}.$$

By analogy with Example 1, Fig. 1.39 prints out a sequence which converges to y with convergence-factor $\frac{1}{16}$; in this case $\sqrt[5]{32}$ is calculated with initial estimate $y_0 = 3$.

```
INP X
Y ← √√X
PRT (Y - 1)/(X - 1)
X ← Y
```

```
INP X
Y ← √√X
PRT 1/(1+Y)(1+YY)
```

Fig. 1.40	Fig. 1.41
0.1892071150	0.1892071150
0.2339963927	0.2339963929
0.2459533709	0.2459533714
0.2489855611	0.2489855660
0.2497462052	0.2497462192
0.2499365115	0.2499365441
0.2499839429	0.2499841354
0.2499962181	0.2499960338
0.2499933816	0.2499990085
0.2499877084	0.2499997521
0.2499130093	0.2499999381
0.2499545977	0.2499999845
0.2492128845	0.2499999961
0.2419825073	0.2499999991
0.2369477912	0.2499999998
0.0000000000	0.2500000000

3. Example

Loss of accuracy through subtraction. Let $x > 0$ and

let $x_0 = x$, $x_1 = \sqrt{\sqrt{x_0}}$, $x_2 = \sqrt{\sqrt{x_1}}$, . . . , $x_n = \sqrt{\sqrt{x_{n-1}}}$ . . .

We know that $\lim_{n \to \infty} x_n = 1$. The convergence-factor equals $\frac{1}{4}$.

Fig. 1.40 shows the sequence $\frac{x_n - 1}{x_{n+1} - 1}$ for $x = 2$ and $n = 1$ to 16.

Up to the eighth member, the sequence approaches the number $\frac{1}{4}$. Then it diverges from the limiting value and the sixteenth number is actually zero. This behaviour is a consequence of severe rounding error. Both x_n and x_{n-1} are close to 1 for $n > 8$. In the expression $x_n - 1$ and $x_{n-1} - 1$ two almost equal numbers are subtracted; thus the earlier reliable figures are lost. This unwelcome effect is called <u>loss of accuracy through subtraction</u>. It can usually be avoided by a re-arrangement

In our case, since $x_{n-1} = x_n^4$, we have

$$\frac{x_n - 1}{x_{n-1} - 1} = \frac{x_n - 1}{x_n^4 - 1} = \frac{x_n - 1}{(x_n^2 + 1)(x_n + 1)(x_n - 1)} = \frac{1}{(1 + x_n^2)(1 + x_n)}$$

Fig. 1.41 shows the astonishing effect of re-arrangement.

Avoid subtracting two almost equal numbers!

Exercises

1. The following expressions are to be evaluated for large values of x.

   a) $\sqrt{x + 1} - \sqrt{x}$, b) $\frac{1}{x + 1} - \frac{1}{x}$, c) $\frac{1}{x + 1} - \frac{2}{x} + \frac{1}{x - 1}$,

   d) $\sqrt{\sqrt{x + 1}} - \sqrt{\sqrt{x}}$.

   Re-arrange them so as to avoid loss of accuracy through subtraction.

2. Write a program to evaluate $y = \sqrt[7]{x}$ by repeated extraction of square roots.

3. Show that $y = \sqrt[9]{x}$ can be obtained by re-arranging the equation $y = \sqrt{\sqrt{\sqrt{\frac{x}{y}}}}$. Use this to write a program for $\sqrt[a]{x}$ and determine $\sqrt[9]{512}$, starting from $y_0 = 1000$. What is the convergence-factor?

4. We consider the sequence in Fig. 1.37

$x_0 = x, \quad x_1 = \sqrt{x}, \quad x_2 = \sqrt{x_1}, \ . \ . \ . \ x_n = \sqrt{x_{n-1}}, \ . \ . \ .$

Because of loss of accuracy through subtraction, $x_n - 1$ cannot be evaluated with high accuracy. Show, however, that

$(x - 1) = (x_n - 1)(x_n + 1)(x_{n-1} + 1)(x_{n-2} + 1) \ldots (x_2 + 1)(x_1 + 1).$

<u>Hint</u>. Notice that $(x_i - 1)(x_i + 1) = (x_i^2 - 1) = (x_{i-1} - 1).$

Write a program which evaluates $x_n - 1$ with high accuracy and check the result $x_{32} - 1 = 1.613859043.10^{-10}$ for $x = 2$.

5. Using (1), it may be shown that

$$y = \sqrt{x} = \left(1 + \frac{1}{q_0}\right)\left(1 + \frac{1}{q_1}\right)\left(1 + \frac{1}{q_2}\right) \ldots = \prod_{n=0}^{\infty} \left(1 + \frac{1}{q_n}\right)$$

where $q_0 = \dfrac{x + 1}{x - 1}, \quad q_{n+1} = 2q_n^2 - 1$

a) Use this to calculate $\sqrt{2}, \ \sqrt{0.5}, \ \sqrt{3}, \ \sqrt{i}$

b) Let $y_0 = 1 + \dfrac{1}{q_0}, \quad y_n = y_{n-1} \cdot \left(1 + \dfrac{1}{q_n}\right)$

Then the approximation y_n has error $\sqrt{x} - y_n \doteq \dfrac{\sqrt{x}}{q_{n+1}}$

Check this for $\sqrt{2}$ and $\sqrt{3}$ using a pocket-calculator.

6. <u>The arithmetico-harmonic mean</u>

Let $x_0 > 0, \quad y_0 > 0$.

We define the sequences $x_n, \ y_n$ by the recursions

$$x_{n+1} = \frac{x_n + y_n}{2}, \quad y_{n+1} = \frac{2x_n y_n}{x_n + y_n}, \quad n = 0, 1, 2, \ . \ . \ .$$

a) Write a program to print (x_n, y_n) for $n = 0, 1, 2, \ . \ . \ .$

b) Experiment with different starting values x_0, y_0 and try to find $\lim_{n \to \infty} x_n$ and $\lim_{n \to \infty} y_n$ in terms of x_0 and y_0.

c) Prove the result you found by experiment in (b).

1.3.5. <u>Logarithms, random digits and exponents</u>

Every real number $a > 0$ can be written as a power of $b > 1$:

1) $a = b^x$

The index x is called the logarithm of a to base b and is written $\log_b a$. For $0 < a < 1$, we have $x < 0$, and for $a \geqslant 1$, $x \geqslant 0$. The most important logarithms are those to base 10, $\log_{10} a = \lg a$; to base 2, $\log_2 a$, and the natural logarithms $\log_e a = \ln a$, where $e = 2.718281828459 0\ldots$ We construct a program which outputs $x = \log_b a$ when a and b are input. We assume that $a > 1$ (the case $0 < a < 1$ is dealt with in Exercise 1). Then $x > 0$.

Let the decimal expression for x be

2) $x = z_0 \; z_1 \, z_2 \, z_3 \ldots = z_0 + \frac{z_1}{10} + \frac{z_2}{100} + \frac{z_3}{1000} + \ldots$

Then $z_0 = [x]$ and z_i is the ith decimal digit of x.

From 1) and 2) it follows that

3) $a = b^{z_0 + z_1/10 + z_2/100 + z_3/1000 + \ldots}$

The following algorithm prints the sequence $z_0, z_1, z_2, z_3, \ldots$

1. Put $z \leftarrow 0$.
2. While $a > b$, put $a \leftarrow \frac{a}{b}$ and $z \leftarrow z + 1$.
3. Print z, put $a \leftarrow a \uparrow 10$ and go to step 1.

i.e. in 3) the factor b is first divided z_0 times into a, and z_0 is printed.

Then 3) becomes

4) $a = b^{z_1/10 + z_2/100 + z_3/1000 + \ldots}$

Now both sides of 4) are raised to the 10th power and we obtain

5) $a = b^{z_1 + z_2/10 + z_3/100 + \ldots}$

This equation has the same form as the initial equation 3).

Fig. 1.42 shows the corresponding program. For $A = 2$, $B = 10$ one obtains

lg2 = 0.30102 99956 62079 68019 75447 64421 62391 57691 78285 99785
72783 97258 38487 31053 95188 60845 79762 84327 96072 57403 . . .

As a consequence of rounding errors, only the first 10 decimal digits are correct. In spite of this the sequence of digits is not without value. If the spinner in Fig. 1.43 was repeatedly spun, a sequence of random decimal digits would be obtained. The digit sequence produced by the program in Fig. 1.42 can serve as a substitute for random decimal digits if $\log_b a$ is irrational. In general one needs at least 1000 random digits, so it is important to speed up the program; for example, one can replace the slow operation $A \leftarrow A \uparrow 10$ by four multiplications:

$P \leftarrow AA, \quad P \leftarrow PP, \quad P \leftarrow PA, \quad A \leftarrow PP.$

The input $A = 2, \; B = e$ gives

$\ln 2 = 0.69314\ 71805\ 46849\ 14692\ 21090\ 10173\ 57756\ 71403 \ldots$

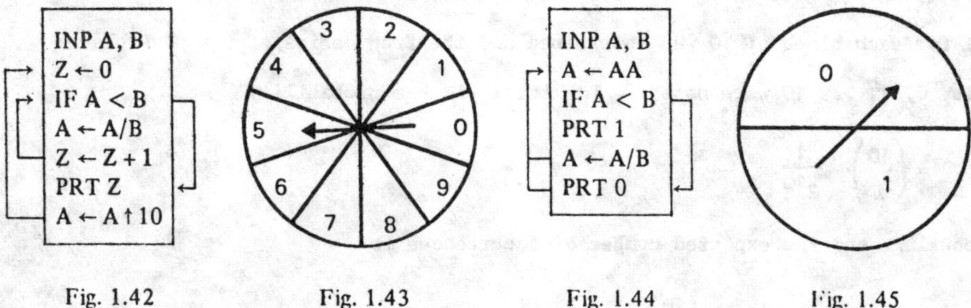

Fig. 1.42 Fig. 1.43 Fig. 1.44 Fig. 1.45

instead of the true value

$\ln 2 = 0.69314\ 71805\ 59945\ 30941\ 72321\ 21458\ 17656\ 80755 \ldots$

This digit sequence is also a good substitute for random decimal digits. It is particularly easy to calculate logarithms in the binary system. The whole number part of the logarithm can be determined mentally, so we shall assume $1 \leq a \leq b$.

Then $\log_b a = (0.z_1\, z_2\, z_3 \ldots)_2 = \frac{z_1}{2} + \frac{z_2}{4} + \frac{z_3}{8} + \ldots, \; z_i \in \{0,1\}$

or $a = b^{z_1/2 + z_2/4 + z_3/8 + \ldots}$

The following algorithm prints the binary sequence $z_1, z_2, z_3, \ldots$

1. (Square) Put $A \leftarrow AA$.
2. If $A < B$, print 0 and go to step 1.

   Otherwise print 1, put $A \leftarrow A/B$ and go to step 1.

The corresponding program in Fig. 1.44 is quickly and easily carried out with a pocket calculator.

$A = 2, \; B = 10$ gives the binary sequence

$$lg2 = 0.0100110100\ 0100000100\ 1101010000\ 1001111101\ 1110010011$$
$$1001110000\ 0010010010\ 1101100001\ 1100011110\ 0111110100$$
$$1001001101\ 0000101001\ 1000110111\ 1101100001\ 0101100101$$
$$0000111011\ 0101000010\ 0100011110\ 0101010010\ 1100011101 \ldots$$

Only the first 36 places are correct.

Throws of a true coin with sides marked 0 and 1, or spins of the spinner shown in Fig. 1.45 produce so-called random binary digits. We expect that Fig. 1.44 gives a very rapid program for production of random binary digits. This belief was subjected to a severe test. The program was used to produce 102400 digits which were then divided into 10240 blocks of 10. The total for each block of 10 was determined and the frequencies of the possible totals 0, 1, ... 10 were noted. The total i has probability

$$p_i = \binom{10}{i} \frac{1}{2^{10}}$$

of occuring, and the expected number of occurrences is

$$E_i = 10240.p_i = 10\binom{10}{i}$$

The program in Fig. 1.46 prints the observed frequencies B_i, which are compared with their expected values E_i. Here A and B are given the values 2 and 5. Lines 30 to 80 generate a block of 10 digits and determine its total S. Lines 20 - 110 generate 10240 such blocks and determine the frequency B(S) of the total S. Lines 120 - 140 print the frequencies in Fig. 1.47, which shows that the values of E_i and B_i are in agreement.

```
10 A = 2
20 FOR  I = 1  TO  10240
30        FOR J = 1  TO 10
40              A = A * A
50              IF A < 5 THEN 80
60              S = S + 1
70              A = 0.2 * A
80        NEXT J
90        B (S) = B (S) + 1
100       S = 0
110 NEXT I
120 FOR  I = 0  TO  10
130       PRINT I, B (I)
140 NEXT I
150 END
```

Fig. 1.46

i	B_i	E_i
0	9	10
1	101	100
2	425	450
3	1235	1200
4	2124	2100
5	2455	2520
6	2149	2100
7	1175	1200
8	466	450
9	91	100
10	10	10

Fig. 1.47

An objective measure of agreement (see [4]) is given by

$$\chi^2 = \sum_{i=0}^{10} \frac{(B_i - E_i)^2}{E_i}$$

The expected value of χ^2 is 1 fewer than the number of rows in Fig. 1.47 i.e. $E(\chi^2) = 10$. In our case, $\chi^2 = 7.51$, which is less than the expected value, so that the agreement is good. Our program has justified itself as a source of random binary digits.

We wish now to calculate b^y for $b > 0$, $y > 0$. If we write $y = n + x$, $n = [y]$, $0 \leqslant x < 1$, then $b^y = b^n . b^x$. Since n is a whole number, the calculation of b^n presents no problems. Thus we can confine ourselves to the case $y = b^x$ with $0 < x < 1$. We write x in binary notation:

$$x = \frac{x_1}{2} + \frac{x_2}{4} + \frac{x_3}{8} + \ldots, \quad x_i \in \{0,1\}$$

$$a = b^x = b^{x_1/2}\, b^{x_2/4}\, b^{x_3/8} \ldots$$

This yields the algorithm in Fig. 1.48; the reader is left to decipher its method of operation.

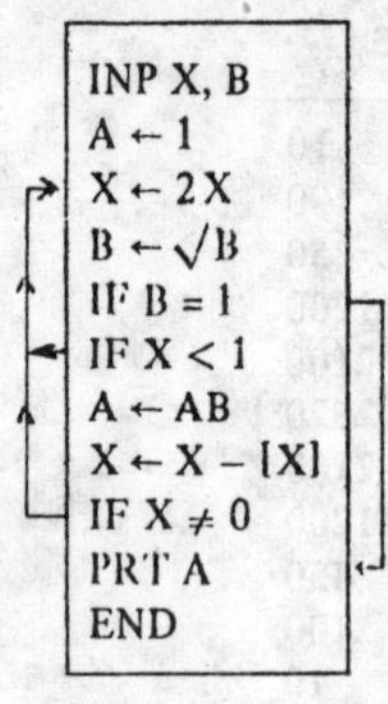

Fig. 1.48

Exercises

1. How can one calculate $\log_b a$, where $0 < a < 1$, using the program of Fig. 1.42?

2. Let $1 < a < 5$. Show that Fig. 1.49 calculates random binary digits.
   Hint: Note that the relation $a \geqslant 5$ takes the value 1 or 0 according to whether it is true or false.

3. Let $1 < a < b$. Fig. 1.50 calculates the digits of $\log_b a$ in the notation using G as base, providing random digits from the set $\{0,1,2 \ldots G - 1\}$

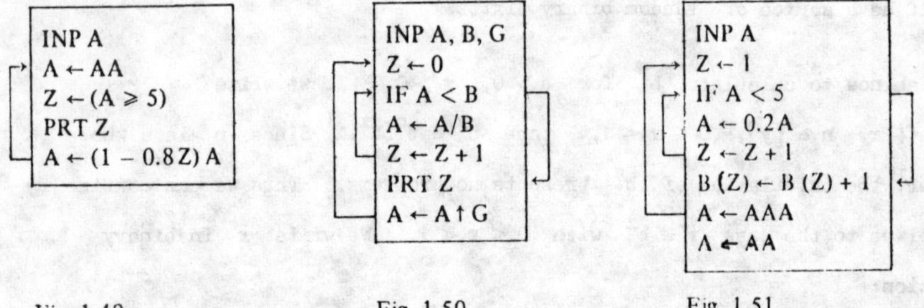

Fig. 1.49 Fig. 1.50 Fig. 1.51

4. From a comparison with Fig. 1.50 it appears that Fig. 1.51 produces throws of a fair dice and counts the frequency of appearance of scores 1 to 6. Write a BASIC program which throws a dice 600 times and prints the frequencies B_i of the outcomes. Test the quality of the random digits using χ^2.

5. Calculate 1000 binary digits of $\log_5 3$, using both Fig. 1.44 and Fig. 1.49;

determine for each the frequency of the digit 1 and compare the times required.

6. Calculate, using Fig. 1.50, 1000 digits of $\log_5 3$ in the octal system (base 8) and determine the frequencies of the digits 0 to 7.

<u>Hint</u>: The program will be quicker if $A \leftarrow A \uparrow 8$ is replaced by $A \leftarrow AA$, $A \leftarrow AA$, $A \leftarrow AA$.

7. Calculate to an accuracy of 10 decimal digits:

$\lg \pi$, $\lg e$, $\ln 10$, $\ln \pi$, $e^{\pi/4}$, $2^{\sqrt{2}}$, π^{π}.

8. The first table of logarithms to base 10 was published in 1617 by Briggs. It is sufficient to provide $\lg x$ for $1 < x < 10$. Briggs, as Napier had done before him, used the formula $\lg \sqrt{ab} = \dfrac{\lg a + \lg b}{2}$. It provides the following algorithm for $\lg x$:

i) Put $a = 1$, $b = 10$, $\lg a = 0$, $\lg b = 1$

ii) Calculate $g = \sqrt{ab}$, $\lg g = \dfrac{\lg a + \lg b}{2}$

iii) If $x < g$, put $b = g$, $\lg b = \lg g$, otherwise put $a = g$, $\lg a = \lg g$.

iv) If $b - a > \varepsilon$, go to step 2, otherwise STOP with answer $\dfrac{\lg a + \lg b}{2}$

Calculate $\lg 2$, $\lg 3$, $\lg 5$ with $\varepsilon = 10^{-10}$ (or with $\varepsilon = 10^{-5}$ if you have only a pocket calculator available).

1.3.6. <u>Random number generators</u>

In 1.3.4. we have already encountered generators for random digits. Almost every computer has its own built-in random number generator. For this we imagine a spinner like that in Fig. 1.52. At the command RND (abbreviation for random) the computer turns the wheel and reads as outcome a real number U in the interval (0, 1). If $0 \leq a \leq b \leq 1$ then U falls in the interval (a, b) with probability $b - a$. The probability U is <u>uniformly distributed</u> in (0, 1). Numbers produced in this way are called <u>random numbers</u> from (0, 1). The BASIC program in Fig. 1.53 prints 10 random numbers.

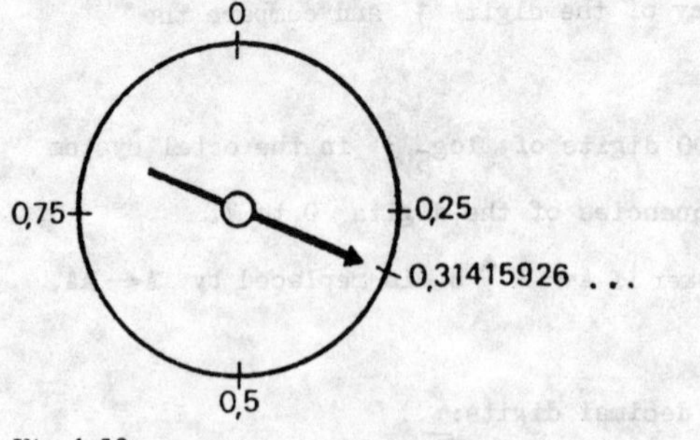

Fig. 1.52

```
10 FOR I = 1 TO 10
20 PRINT RND
30 NEXT I
40 END
```

Fig. 1.53

We wish to choose a number <u>at random</u> from $\Omega = \{0, 1, 2, \dots N - 1\}$ i.e. so to choose that each element of Ω has probability $\frac{1}{N}$ of being chosen. A number chosen in this way is called a <u>random digit</u> from Ω. The instruction N. RND chooses a (real) random number from (0, N). Hence the instruction [N. RND] gives a random digit from Ω and [N. RAND] + 1 gives a random digit from $\{1, 2, 3, \dots N\}$. In BASIC, INT (2*RND) simulates one throw of a true coin with sides labelled 0 and 1. INT(6*RND) + 1 simulates a throw of a fair die, and INT (10*RND) gives a random decimal digit. In this way one can easily produce random digits from random numbers. Many desk-calculators have no in-built random number generator, so we shall discuss a few algorithms for generation of random numbers.

a) The mid-square method (J. von Neumann, 1946): Start with an n-digit natural number, A, square it and select the middle n digits of the square as the next number. A sequence of natural numbers from $\{0, 1, 2, \dots 10^n - 1\}$ is obtained; division by 10^n gives numbers from (0, 1) which may be used in place of random numbers. Fig. 1.54 uses $n = 4$. This was historically the first attempt to create artificial random numbers by a deterministic process. Unfortunately this simple method degenerates very quickly. <u>It should not be employed.</u> (See Exercise 3).

5. Verify that the Fibonacci sequence modulo m

$X_1 = X_2 = 1, \; X_{n+1} \equiv X_n + X_{n-1} \pmod{m}$ produces random digits of very poor quality (see 1.4).

6. In Fig. 1.56 let Z be a real number chosen from (0, 1)

e.g. $\frac{\pi}{4}$, $\frac{\sqrt{2}}{2}$, $\frac{\sqrt{5}-1}{2}$. The program produces a sequence of real numbers from (0, 1). The fractional part of the reciprocal of one member of the sequence is the next member of the sequence. Use it to produce 1000 random decimal digits and determine the fequency of the digits 0 to 9. Test several different starting values. Why is the starting value $\frac{\sqrt{5}-1}{2}$ particularly bad? What is noteworthy about the starting value $\frac{\sqrt{2}}{2}$?

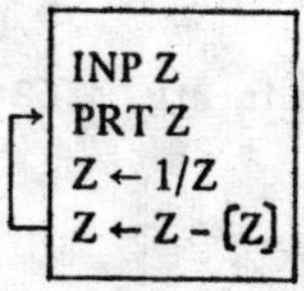

```
   INP Z
┌→ PRT Z
│  Z ← 1/Z
└─ Z ← Z - [Z]
```

Fig. 1.57

1.4 The Fibonacci sequence

Leonardo of Pisa, called Fibonacci (i.e. son of Bonacci) was the greates mathematician of the Middle Ages. In 1202 his book "liber abaci" appeared in which the Hindu-Arabic digits were introduced to Europe. In this book is also the famous Rabbit Problem:

A pair of rabbits is born at time 0 and produces a further pair in the second month of its existence and in each subsequent month. Each succeeding generation reproduces in the same way. How many pairs are there after n months? Table 1.58 shows the total number of pairs in the first year.

Time	0	1	2	3	4	5	6	7	8	9	10	11	12
Pairs	1	1	2	3	5	8	13	21	34	55	89	144	233

Fig. 1.58

Nowadays we define the Fibonacci sequence by

(1) $F_1 = F_2 = 1, \quad F_{n+2} = F_n + F_{n+1}, \quad n \geqslant 0$

This sequence has a multitude of interesting properties which have been examined in several thousand articles. There is even a Fibonacci Society and a magazine "Fibonacci Quarterly". We wish to print the sequence $F_1, F_2, F_3, \ldots$ The two most recently printed members we shall call A and B. Initially A = B = 1. After each member is output the pair (A, B) is replaced by (B, A + B). Before the substitution A← B we make a copy C of A, since we need the old value of A to calculate the next member A + B. Running through the steps of the program in Fig. 1.59 establishes that it prints out the sequence F_n. The copy C is superfluous as the programs in Fig. 1.60 and 1.61 show. The latter is particularly economical, since it calculates two members using the same number of instructions.

As a rule we shall use Fig. 1.60. The BASIC program in Fig. 1.62 prints F_1 to F_{60}.

```
A ← B ← 1
→ PRT A
  C ← A
  A ← B
  B ← B + C
```

Fig. 1.59

```
A ← B ← 1
→ PRT A
  B ← A + B
  A ← B - A
```

Fig. 1.60

```
A ← B ← 1
→ PRT A, B
  A ← A + B
  B ← A + B
```

Fig. 1.61

We consider now the sequence of quotients $q_n = \dfrac{F_{n+1}}{F_n}$

$\frac{1}{1}, \frac{2}{1}, \frac{3}{2}, \frac{5}{3}, \frac{8}{5}, \frac{13}{8}, \frac{21}{13}, \ldots$

```
10 A = B = 1
20 FOR I = 1 TO 30
30      PRINT A, B,
40      A = A + B
50      B = A + B
60 NEXT I
70 END
```

Fig. 1.62

1	1	2	3	5
8	13	21	34	55
89	144	233	377	610
987	1597	2584	4181	6765
10946	17711	28657	46368	75025
121393	196418	317811	514229	832040
1346269	2178309	3524578	5702887	9227465
14930352	24157817	39088169	63245986	102334155
165580141	267914296	433494437	701408733	1134903170
1836311903	2971215073	4807526976	7778742049	12586269025
20365011074	32951280099	53316291173	86267571272	139583862445
225851433717	365435296162	591286729879	956722026041	1548008755920

Fig. 1.62

The program in Fig. 1.63 prints 25 members of this sequence. The table suggests the conjecture that q_n tends to the limit

$$\phi = \lim_{n \to \infty} q_n = 1.6180339887 \ldots$$

The existence of the limit will be proved in Exercise 5.

It is easy to deduce an equation for ϕ

$$q_n = \frac{F_{n+1}}{F_n} = \frac{F_n + F_{n-1}}{F_n} = 1 + \frac{F_{n-1}}{F_n} = 1 + \frac{1}{q_{n-1}}$$

Since $\lim_{n \to \infty} q_n = \lim_{n \to \infty} q_{n-1} = \phi$, it follows that , for $n \to \infty$,

$$\phi = 1 + \frac{1}{\phi}, \quad \phi = \frac{1 + \sqrt{5}}{2} = 1.61803398874989484820\ldots$$

ϕ is the well-known ratio of the golden section

```
10 A = B = 1
20 FOR I = 1 TO 30
30       PRINT B/A,
40       B = A + B
50       A = B − A
60 NEXT I
70 END
```

1	2	1.5	1.6666666667	1.6
1.625	1.6153846154	1.6190476190	1.6176470588	1.6181818182
1.6179775281	1.6180555556	1.6180257511	1.6180371353	1.6180327869
1.6180344478	1.6180338134	1.6180340557	1.6180339632	1.6180339985
1.6180339850	1.6180339902	1.6180339882	1.6180339890	1.6180339887
1.6180339888	1.6180339887	1.6180339888	1.6180339887	1.6180339887

Fig. 1.63

We now examine the Fibonacci sequence modulo 10:

2) 1, 1, 2, 3, 5, 8, 3, 1, 4, 5, 9, 4, 3, 7, 0, 7, 7, 4, 1, 5, 6, 6, 1, 7, 8, 5, 3, 8, 1, 9, 0...

We show that (2) is periodic from the beginning. Each pair of neighbouring members determines all the following, and preceding, members. The pair (0, 0) cannot occur (why not?). So there are altogether 99 possible pairs (0, 1), (0, 2). . . (9, 9). 101 consecutive members of (2) give 100 neighbouring pairs. Therefore at least one pair must be repeated:

$$(F_i, F_{i+1}) = (F_k, F_{k+1}), \quad 1 \leq i < k \leq 100.$$

Then all preceding pairs must agree and are similarly determined:

$$(F_{i-t}, F_{i+1-t}) = (F_{k-t}, F_{k+1-t}), \; t = 1, 2, 3, \ldots i - 1$$

i.e. the sequence is periodic, the period begins with 1, 1... and has length k - 1. In an exactly similar way one may show that the Fibonacci sequence is immediately periodic for every modulus M. The program in Fig. 1.64 prints one period of the Fibonacci sequence modulo M, and the period-length P. For M = 10 the result is P = 60. We want to find a connection between M and P. So we print a table showing, for each modulus M from 2 to 125, the corresponding value of P. The outcome (Fig. 1.65) is examined in Exercise 1.

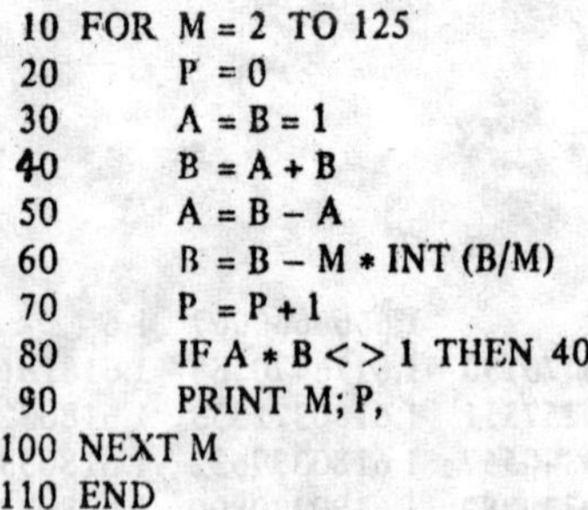

```
10 FOR M = 2 TO 125
20     P = 0
30     A = B = 1
40     B = A + B
50     A = B - A
60     B = B - M * INT (B/M)
70     P = P + 1
80     IF A * B < > 1 THEN 40
90     PRINT M; P,
100 NEXT M
110 END
```

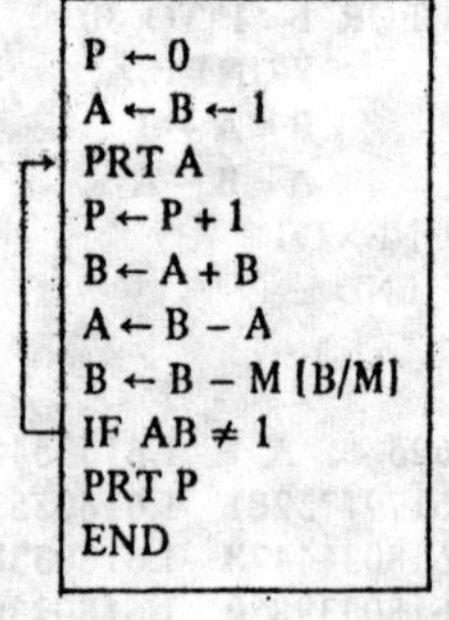

Fig. 1.64

2	3	3	8	4	6	5	20	6	24
7	16	8	12	9	24	10	60	11	10
12	24	13	28	14	48	15	40	16	24
17	36	18	24	19	18	20	60	21	16
22	30	23	48	24	24	25	100	26	84
27	72	28	48	29	14	30	120	31	30
32	48	33	40	34	36	35	80	36	24
37	76	38	18	39	56	40	60	41	40
42	48	43	88	44	30	45	120	46	48
47	32	48	24	49	112	50	300	51	72
52	84	53	108	54	72	55	20	56	48
57	72	58	42	59	58	60	120	61	60
62	30	63	48	64	96	65	140	66	120
67	136	68	36	69	48	70	240	71	70
72	24	73	148	74	228	75	200	76	18
77	80	78	168	79	78	80	120	81	216
82	120	83	168	84	48	85	180	86	264
87	56	88	60	89	44	90	120	91	112
92	48	93	120	94	96	95	180	96	48
97	196	98	336	99	120	100	300	101	50
102	72	103	208	104	84	105	80	106	108
107	72	108	72	109	108	110	60	111	152
112	48	113	76	114	72	115	240	116	42
117	168	118	174	119	144	120	120	121	110
122	60	123	40	124	30	125	500		

Fig. 1.65

Exercises

1. Let $L(m)$ be the period-length of the Fibonacci sequence modulo m. Verify the following conjectures using Table 1.65.

   a) $L(p^n) = L(p)\ p^{n-1}$ for each prime number p.

   b) $L(p_1^{\alpha_1}\ p_2^{\alpha_2} \ldots p_r^{\alpha_r}) = \mathrm{LCM}\left\{ L(p_1^{\alpha_1}), L(p_2^{\alpha_2}), \ldots L(p_r^{\alpha_r}) \right\}$ for primes $p_1, p_2 \ldots p_r$.

2. Use Exercise 1 to determine $L(10^n)$.

3. To find the first member of the Fibonacci sequence F_n which ends in m zeros, one must determine the first occurrence of 0 in the sequence F_n modulo 10^m. Write a program which prints the number of the first Fibonacci number divisible by 10^m. What does one obtain for $m = 1, 2, 3, 4$?

4. We wish to examine the divisibility properties of the Fibonacci sequence. Write a BASIC program, which prints in separate rows the subscript

numbers of those Fibonacci numbers up to F_{50}, which are divisible by 2, 3, ... 20. State some conjectures about divisibility properties.

```
    A ← B ← 1
┌→  PRT A
│   A ↔ B
└   B ← A + B
```

Fig. 1.66

5. Let $\phi = \frac{\sqrt{5}+1}{2}$ and $q_n = \frac{F_{n+1}}{F_n}$. Show by induction that

$q_n - \phi = \frac{(-1)^n}{F_n \phi^n}$. From this it follows that $\lim_{n\to\infty} q_n = \phi$.

6. Some computers possess an exchange instruction ⟷ , i.e. A⟷B interchanges the contents of the stores A and B. Verify that the program in Fig. 1.66 prints the Fibonacci sequence.

7. Let A, B, C be three consecutive members of the Fibonacci sequence. Write a program which prints $B^2 - AC$. Conjecture? Proof?

8. The <u>Lucas sequence</u> is defined by $L_1 = 2$, $L_2 = 1$, $L_{n+2} = L_n + L_{n+1}$, $n \geqslant 0$. It is closely related to the Fibonacci sequence.

   a) Let A, B, C be three consecutive members of the Lucas sequence. Write a program to print $B^2 - AC$. Conjecture? Proof?

   Investigate the value of $p_n = \frac{L_{n+1}}{L_n}$ for $n\to\infty$, and the period-length of the Lucas sequence modulo M for M = 2 to 100.

[9]. We consider the sequence of first digits of F_1 to F_{10000}

1, 1, 2, 3, 5, 8, 1, 2, 3, 5, 8, 1, 2, 3, 6, 9, 1, 2, 4, 6, 1, 1, 2, 4, 7...

Write a program which determines the frequencies of the digits 1 to 9 in the sequence.

2. NUMBER THEORY

2.1. Conversion from base 10 to base b and vice versa.

a) Conversion of natural numbers from base 10 to base b.

Let $b \geqslant 2$ be a natural number. Then every natural number z can be uniquely expressed as a sum of powers of b with coefficients $a_i \in \{0, 1, 2, \ldots b-1\}$:

$$z = a_n b^n + a_{n-1} b^{n-1} + \ldots + a_2 b^2 + a_1 b + a_0 = (a_n a_{n-1} \ldots a_2 a_1 a_0)_b$$

The coefficients a_i are called digits. If z is divided by b, the quotient and remainder are given by

$$q = \left[\frac{z}{b}\right] = a_n b^{n-1} + a_{n-1} b^{n-2} + \ldots a_2 b + a_1, \quad r = z - bq = z - b\left[\frac{z}{b}\right] = a_0$$

If we replace z by q and divide again by b, the remainder will be the next digit a_1, and so on. i.e. Fig. 2.1 prints the digits in the sequence $a_0 \ldots a_n$. The program in Fig. 2.2 is one line longer, and needs one more store but saves one division.

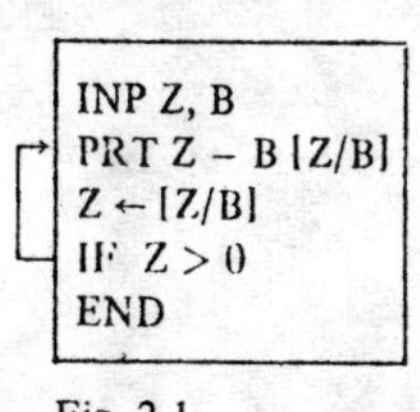

Fig. 2.1

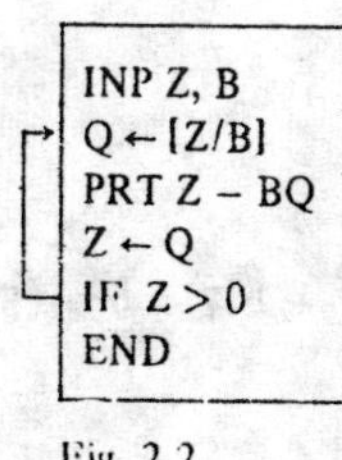

Fig. 2.2

```
  5 DIM R (40)
 10 READ B, Z
 20 FOR I = 1 TO 40
 30     R (I) = Z – B*INT (Z/B)
 40     Z = INT (Z/B)
 50     IF Z = 0 THEN 70
 60 NEXT I
 70 FOR J = I TO 1 STEP – 1
 80     PRINT R (J);
 90 NEXT J
100 PRINT
110 GOTO 10
120 DATA 2, 1976, 3, 123456, 5,8192
130 END

1 1 1 1 0 1 1 1 0 0 0
2 0 0 2 1 1 0 0 1 1 0
2 3 0 2 3 2
```

Fig. 2.3

The BASIC program in Fig. 2.3 prints the digits in the correct sequence $a_n, a_{n-1}, \ldots a_0$. We discuss it, since it contains a new instruction. Line number 5 reserves the stores R(0) to R(40), in which we will place the digits. The computer will automatically reserve the stores

R(0) to R(10). If more space is needed, a <u>dimension statement</u> must be used to reserve it. It is allowable to reserve more space than will be needed.

When the computer reaches the READ instruction (line 10) it goes to the DATA statement (line 120), and sets B = 2, z = 1976. In 20-60 the digits of z in the base B are calculated and a_{i-1} is stored in R(I). 70-90 prints the digits in the correct sequence, beginning with a_n, 100 prints a new line, 110 returns control to line 10. From there the computer goes to 120, sets B = 3 and z = 123456 and so on ... After it has calculated 8192 in base 5, it runs out of data and the program stops.

If it is required to run the program with fresh data, it is only necessary to change line 120.

b) <u>Conversion of a natural number in base b to base 10</u>

If we have a natural number z in base b

$$z = a_n b^n + a_{n-1} b^{n+1} + \dots a_1 b + a_0 ,$$

we calculate its value in base 10 by substituting the value of b in the polynomial

$$f(x) = a_n x^n + a_{n-1} x^{n-1} + \quad + a_1 x + a_0 .$$

There is an elegant algorithm for this, given by Newton and known as the "<u>Horner method</u>". Using brackets, we obtain

$$f(x) = (. . .(((a_n)x + a_{n-1})x + a_{n-2})x + . . . a_1)x + a_0$$

We calculate as the brackets tell us to:

	<u>Example:</u> $z = 5.7^4 + 2.7^3 + 3.7^2 + 4.7 + 6 = (52346)_7$
$z \leftarrow 0$	$z \leftarrow 0$
$z \leftarrow zx + a_n$	$z \leftarrow 0.7 + 5 = 5$
$z \leftarrow zx + a_{n-1}$	$z \leftarrow 5.7 + 2 = 37$
$z \leftarrow zx + a_{n-2}$	$z \leftarrow 37.7 + 3 = 262$
....	$z \leftarrow 262.7 + 4 = 1838$
$z \leftarrow zx + a_n$	$z \leftarrow 1838.7 + 6 = 12872$

Fig. 2.4 shows a BASIC program for the Horner method.

```
 10 DIM R (40)
 20 READ X, N
 30 FOR I = 0  TO  N
 40         READ R (I)
 50 NEXT I
 60 Z = 0
 70 FOR  I = 0  TO  N
 80         Z = Z*X + R (I)
 90 NEXT I
100 PRINT Z
110 GOTO  20
120 DATA  7,4,5,2,3,4,6
130 DATA  8,6,2,0,3,0,4,2,1
140 END

    12872
    536849
```

Fig. 2.4

Commentary.

10 The stores R(0) to R(40) are reserved for the coefficients $a_n, a_{n-1}, \ldots a_0$

20 The base X = 7 and the degree of the polynomial N = 4 are read in

30-50 The co-efficients R(0) = 5, R(1) = 2, R(2) = 3, R(3) = 4, R(4) = 6, are read in

60-90 The polynomial $f(x) = 5x^4 + 2x^3 + 3x^2 + 4x + 6$ is evaluated at the point $x = 7$

100 $f(x)$ is printed

110 Data for the next exercise are fetched.

c) <u>Conversion of a number $z < 1$ from base 10 to base b</u>

We must express z in powers of b^{-1}. In general the expression will be non-terminating. So we choose the number of places, n, at which we break off the expression.

$$z = \frac{a_{-1}}{b} + \frac{a_{-2}}{b^2} + \frac{a_{-3}}{b^3} + \ldots = (0, a_{-1}, a_{-2}, a_{-3}, \ldots)_b$$

Multiplication by b gives

$$bz = a_{-1} + \frac{a_{-2}}{b} + \frac{a_{-3}}{b} + \quad \ldots = (a_{-1}, a_{-2}, a_{-3}, \ldots)_b$$

$$a_{-1} = [bz], \quad r = bz - [bz] = (0, a_{-2}, a_{-3}, \ldots)_b$$

Replacing z by r and repeating this step, we obtain the successive digits $a_{-1}, a_{-2}, a_{-3}, \ldots$ The program in Fig. 2.5 calculates n places in base b.

```
   INP B, Z, N
   I ← 1
┌→ P ← BZ
│  PRT [P]
│  Z ← P − [P]
│  I ← I + 1
└─ IF I ≤ N
   END
```

Fig. 2.5

d) Conversion of a number $z < 1$ from base b to base 10

If z is an infinite expression in base b, we take the first n places:

$$z = (0, a_{-1}, a_{-2}, \ldots a_{-n})_b = \frac{a_{-1}}{b} + \frac{a_{-2}}{b^2} + \ldots \frac{a_{-n}}{b^n}$$

This polynomial in $c = \frac{1}{b}$ may be evaluated by Horner's method.

Exercises

1. Fig. 2.5a gives a simplified version of Fig. 2.5. The program prints an infinite sequence of digits. Verify the following using this program:

   $\frac{1}{3} = 0.\overline{01}_2$, $\frac{1}{5} = 0.\overline{0011}_2$, $\frac{2}{7} = 0.\overline{010}_2$, $\frac{3}{4} = 0.\overline{20}_3$, $\frac{3}{5} = 0.\overline{1210}_3$,

   $\frac{1}{7} = 0.\overline{1}_8$, $\frac{1}{7} = 0.\overline{142857}_{10}$, $\frac{6}{13} = 0.\overline{461538}_{10}$.

   (The period is marked by overlining in each case)

   How many correct places are given by the computer?

2. Let Z and B be two natural numbers, $B \geq 2$. Write a program, which, when Z and B are input, gives the digital root (sum of digits) of Z in base B.

3. We describe an algorithm for production of a number sequence :

   1) Start with a natural number A, which is a multiple of 3.

   2) Print A. If A = 153, the STOP.

   3) Otherwise replace A by the sum of the cubes of the digits of A

and go to step 2.

a) Write a program which prints the corresponding sequence when a value A is input.

[b] Prove, with the help of the computer, that the algorithm always stops.

► 4. (cf. Fig. 1.30). We wish to calculate $z = x^y$, where $y (\geqslant 0)$ is a whole number

e.g. $z = x^{90} = x^{1011010_2} = x^{64} . x^{16} . x^8 . x^2$.

If the instructions $\begin{cases} x \leftarrow x\,x \\ y \leftarrow \left[\frac{y}{2}\right] \end{cases}$ are performed while $y \neq 0$,

x and y respectively run through the values x, $\underline{x^2}$, x^4, $\underline{x^{16}}$, x^{32}, $\underline{x^{64}}$ and 90, <u>45</u>, 22, <u>11</u>, <u>5</u>, 2, <u>1</u>. $z = x^{90}$ is the product of all powers which correspond to odd values of y. The program in Fig. 2.5b is based on this. Run it through a few times by hand. In calculating x^y, the squaring is performed once too often. Correct this "defect".

(cf. Fig. 1.14a).

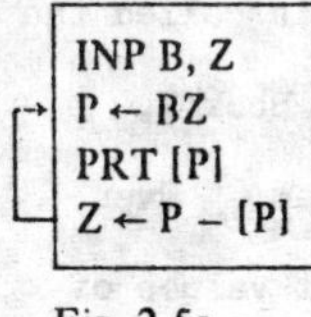

Fig. 2.5a

```
10 INPUT X, Y
20 Z = 1
30 IF Y/2 = INT (Y/2) THEN 50
40 Z = Z * X
50 X = X * X
60 Y = INT (Y/2)
70 IF Y <> 0 THEN 30
80 PRINT Z
90 END
```

Fig. 2.5b

2.2. The Euclidean algorithm

Let a and b be two non-negative whole numbers. Their <u>highest common factor</u> (hcf) we denote by $a \sqcap b$. If we write $0 \sqcap 0 = 0$, then the following hold for all $a, b \in \{0, 1, 2, \ldots\}$.

$a \sqcap 1 = 1$, $a \sqcap a$, $a \sqcap 0 = a$, $a \sqcap b = b \sqcap a$

e.g. $10 \sqcap 15 = 5$, $9 \sqcap 10 = 1$, $7 \sqcap 7 = 7$, $8 \sqcap 0 = 8$, $6 \sqcap 1 = 1$.

Let $a \geqslant b \geqslant 0$. By division we have

1) $a = bq + r, \quad 0 \leq r < b$

2) $q = \left[\frac{a}{b}\right], \quad r = a - bq = a - b\left[\frac{a}{b}\right] = a \bmod b$

From 1) it follows easily that $a \sqcap b = b \sqcap r$

Thus we can replace the pair (a, b) by the smaller pair (b, r) with the same hcf. Repeating this step, we obtain ever smaller pairs, until finally a pair (h, 0) is obtained. Then $a \sqcap b = h \sqcap 0 = h$.

We illustrate with the numerical examples $560 \sqcap 91$ and $972 \sqcap 666$:

$560 = 91.6 + 14$	$972 = 666.1 + 306$
$91 = 14.6 + 7$	$666 = 306.2 + 54$
$14 = 7.2 + 0$	$306 = 54.5 + 36$
	$54 = 36.1 + 18$
	$36 = 18.2 + 0$

i.e. $560 \sqcap 91 = 7 \sqcap 0 = 7$ and $972 \sqcap 666 = 18 \sqcap 0 = 18$.

This elegant and rapid algorithm for determination of the hcf is called the Euclidean algorithm. It is found in Euclid's "Elements", Book 7, Propositions 1 and 2. **Fig.** 2.6 shows the corresponding program. Even more elegant is the program of Fig. 2.7. Try it for different values of a and b.

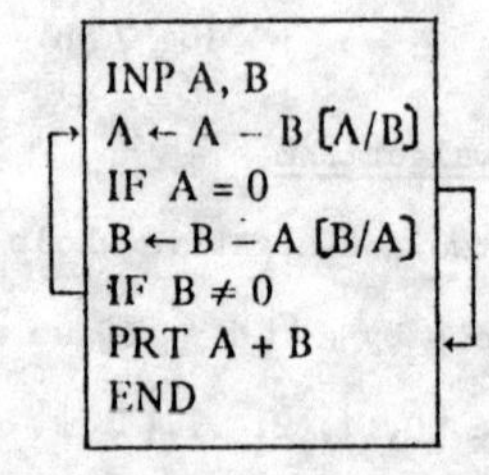

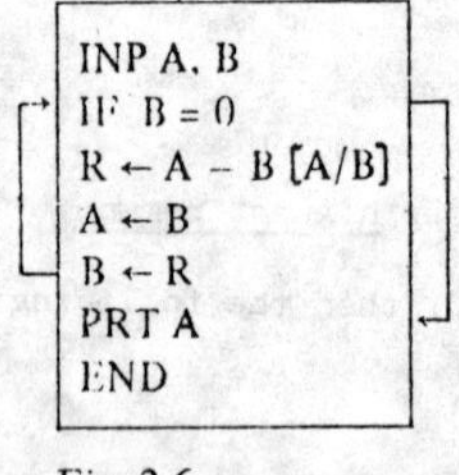

Fig. 2.6

```
INP A, B
A ← A – B [A/B]
IF A = 0
B ← B – A [B/A]
IF B ≠ 0
PRT A + B
END
```

Fig. 2.7

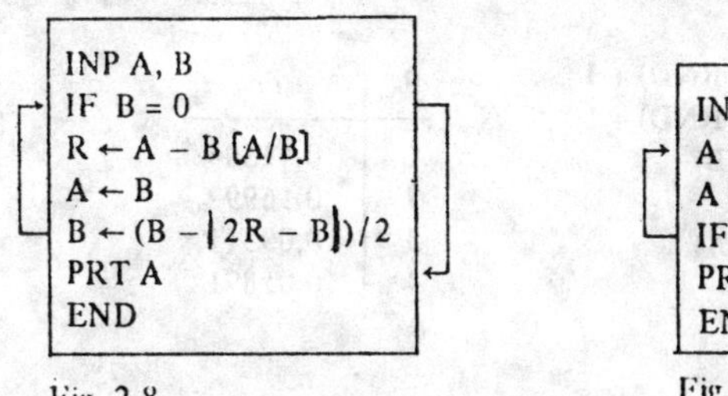

Fig. 2.8

```
INP A, B
A ← A mod B
A ⟷ B
IF B ≠ 0
PRT A
END
```

Fig. 2.9

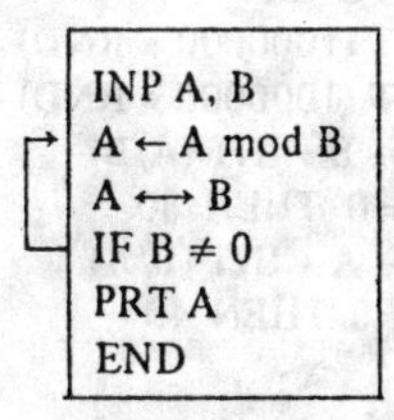

By using the mod operation and the exchange instruction one obtains Fig. 2.9. Check the correctness of this program. We wish to shorten the Euclidean algorithm even more.

In the division $a = bq + r,\ \ 0 \leq r < b$

we have used a positive remainder r. We can also work with negative remainders.

Clearly $a = bq + r = b(q + 1) - (b - r),\ 0 \leq b - r < b,$

and $a \sqcap b = b \sqcap r = b \sqcap (b - r)$

If $b - r < r$, then it is better to use the replacement $b \leftarrow b - r$ instead of $b \leftarrow r$, in Fig. 2.6, i.e. the replacement used is $b \leftarrow \min(r, b - r)$.

From 1.3.1, Exercise 1, we have

$$\min(r, b - r) = \frac{r + b - r - |r - b + r|}{2} = \frac{b - |2r - b|}{2}$$

From this we obtain the <u>Euclidean algorithm with least absolute remainder</u> in Fig. 2.8.

We calculate $89 \sqcap 55$ using Figs. 2.6 and 2.8.

$89 \sqcap 55 = 55 \sqcap 34 = 34 \sqcap 21 = 21 \sqcap 13 = 13 \sqcap 8 = 8 \sqcap 5 = 5 \sqcap 3 = 3 \sqcap 2 = 2 \sqcap 1 = 1$

$89 \sqcap 55 = 55 \sqcap 21 = 21 \sqcap 8 = 8 \sqcap 3 = 3 \sqcap 1 = 1$

The new algorithm requires fewer, though more complicated, steps. To determine which of the three algorithms is the best, we took 1000 pairs (a, b) and found the computation time. The numbers a and b were randomly chosen from $\{1, 2, \ldots 1000000\}$. Fig. 2.10 shows the BASIC program corresponding to Fig. 2.7. In the contest Fig. 2.7 is the clear winner. Fig. 2.8 is next best and Fig. 2.6 the slowest. From now on we shall regard Fig. 2.7 as our standard program.

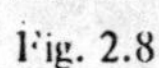

```
10 FOR I = 1  TO 1000
20        A = INT (1000000 * RND) + 1
30        B = INT (1000000 * RND) + 1
40        A = A - B * INT (A/B)
50        IF  A = 0  THEN  80
60        B = B - A * INT (B/A)
70        IF  B ≠ 0  THEN  40
80 NEXT I
90 END
```

Fig. 2.10

n	p_n
1	0,41504
2	0,16992
3	0,09311
4	0,05891

Fig. 2.11

Nicomachus (100 AD) produced a version of the algorithm which depends on subtraction rather than division. It used the fact that

$a \sqcap b = b \sqcap (a - b),\ a \geq b.$

Examples: $120 \sqcap 48 = 72 \sqcap 48 = 48 \sqcap 24 = 24 \sqcap 24 = 24$

$8 \sqcap 5 = 5 \sqcap 3 = 3 \sqcap 2 = 2 \sqcap 1 = 1$

i.e. the smaller number is always subtracted from the larger, until one of the numbers becomes 1 or both are equal. The process appears wasteful. It can be shown that in the division the quotient n occurs with probability

$$p_n = \log_2 \frac{(n+1)^2}{(n+1)^2 - 1} \qquad \text{(Fig. 2.11)}$$

i.e. in 58.5% of all cases, a relatively expensive division can be replaced by one or two subtractions.

There follows a description of the Nicomachus algorithm:

1. If $a = 1$ or $b = 1$ then STOP, with 1 as answer
2. If $a = b$ then STOP with a as answer
3. If $a > b$, let $a \leftarrow a - b$ and go to step 1
4. If $a < b$, let $b \leftarrow b - a$, and go to step 1.

Exercises

1. Check that the programs in Figs. 2.6 - 2.9 are also correct for $a < b$.
2. Why is $b = 0$ not allowed in Figs. 2.7 and 2.9?
3. Draw a flow diagram for the Nicomachus algorithm.
4. The programs in Figs. 2.6 - 2.8 and the Nicomachus algorithm are to be compared.

   a) Write a program analogous to Fig. 2.10 and compare the computation times.

b) Choose a, b from $\{1, 2, \ldots, 1000\}$ and compare again.

5. Choose two numbers a, b randomly from $\{1, 2, \ldots, 10^6\}$ and find the quotient of the larger by smaller. Repeat this 1000 times and count how often $[q] = 1, 2, 3, 4$ arises. Compare with Fig. 2.11.

6. It can be shown that, if a, b are randomly chosen from $\{1, 2, \ldots, n\}$ the mean number of divisions needed to determine $a \sqcap b$ is $\frac{12 \ln(2)}{\pi^2} \ln(n) + 0.06$. For $n = 10^6$ this gives 11. 703 divisions. This can be checked by incorporating in Fig. 2.10 a number D which counts the divisions.

▶ 7. Two numbers are chosen randomly from $\{1, 2, \ldots 10^6\}$. If $a \sqcap b = 1$, then Abel wins, otherwise Cain wins. Write a program to play the game 1000 times and count Abel's wins. Is the game fair?

▶ 8. A triple of natural numbers is a Pythagorean triple if $x^2 + y^2 = z^2$. The triple is primitive if $x \sqcap y \sqcap z = 1$. The best-known such triples are (3, 4, 5) and (5, 12, 13). We require a table of primitive Pythagorean triples. They can all be obtained from

$x = a^2 - b^2$, $y = 2ab$, $z = a^2 + b^2$

where 1) $a > b$, 2) $a \sqcap b = 1$, 3) a and b have opposite parity. (Two natural numbers have opposite parity when one is even and the other is odd). Write a BASIC program to print all Pythagorean triples with $a \leq 10$.

9. The lowest common multiple of the natural numbers a and b is denoted by $a \sqcup b$. Find algorithms for determining $a \sqcup b$.

2.3.* An extension of the Euclidean algorithm

If a and b are natural numbers with highest common factor d, there are integers x, y such that

1) $ax + by = d$

i.e. the hcf of a and b can be written as a linear combination of a and b with integer coefficients. In almost all applications x and y are needed, as well as d. We give an algorithm which for given a, b, calculates d, x and y in 1). For this we introduce a new notation. We write $a_0 = a$, $a_1 = b$ and use the Euclidean algorithm to find the sequence of remainders $a_0, a_1, a_2, \dots a_k$, where a_i is the remainder when a_{i-2} is divided by a_{i-1}. Here a_k is the last non-zero remainder – i.e. a_k divides a_{k-1} and $a_{k+1} = 0$. Then $a_0 \sqcap a_1 = a_k$. With the remainder sequence we also calculate two further sequences x_i, y_i. We show that the algorithm in Fig. 2.12 prints the desired d, x, y of 1). In order to understand it, we go through the process for three numerical examples.

a) $a_0 = 91$, $a_1 = 56$ b) $a_0 = 286$, $a_1 = 121$ c) $a_0 = 119$, $a_1 = 13$.

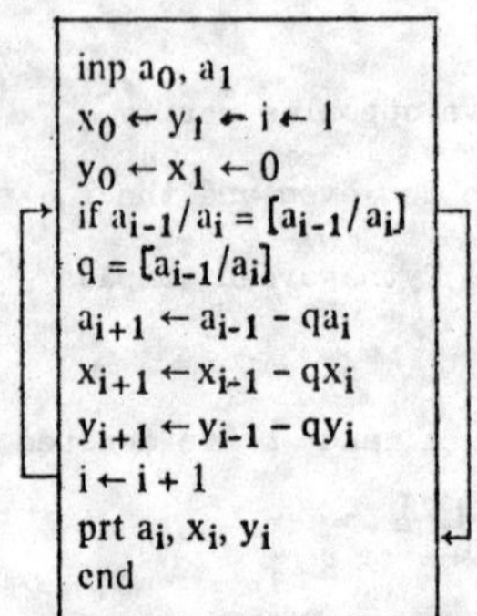

```
inp a0, a1
x0 ← y1 ← i ← 1
y0 ← x1 ← 0
if a(i-1)/ai = [a(i-1)/ai]
q = [a(i-1)/ai]
a(i+1) ← a(i-1) - q·ai
x(i+1) ← x(i-1) - q·xi
y(i+1) ← y(i-1) - q·yi
i ← i + 1
prt ai, xi, yi
end
```

Fig. 2.12

i	a_i	x_i	y_i
0	91	1	0
1	56	0	1
2	35	1	-1
3	21	-1	2
4	14	2	-3
5	7	-3	5

$7 = -3 \cdot 91 + 5 \cdot 56$

i	a_i	x_i	y_i
0	286	1	0
1	121	0	1
2	44	1	-2
3	33	-2	5
4	11	3	-7

$11 = 3 \cdot 286 - 7 \cdot 121$

i	a_i	x_i	y_i
0	119	1	0
1	13	0	1
2	2	1	-9
3	1	-6	55

$1 = -6 \cdot 119 + 55 \cdot 13$

We begin with rows 0 and 1: $a_0 \quad 1 \quad 0$

$a_1 \quad 0 \quad 1$

Row $(i + 1)$ is obtained from rows $(i - 1)$ and i as follows:

The value $q = \left[\frac{a_{i-1}}{a_i}\right]$ is determined. Then from the $(i - 1)$th row is subtracted q times the ith row.

We prove the correctness of the algorithm. We show that, for $i \geqslant 0$

2) $a_0x_i + a_1y_i = a_i$

The equation is correct for $i = 0$ and $i = 1$.

For $\quad a_0 .1 + a_1 .0 = a_0$

$a_0 .0 + a_1 .1 = a_1$

We assume that 2) is correct for $i - 1$ and for i. Since

$x_{i+1} = x_{i-1} - qx_i$, $y_{i+1} = y_{i-1} - qy_i$, $a_{i+1} = a_{i-1} - qa_i$

and using the induction hypothesis, it follows that

$$a_0x_{i+1} + a_1y_{i+1} = a_0x_{i-1} + a_1y_{i-1} - q(a_0x_i + a_1y_i)$$

$$= a_{i-1} - qa_i = a_{i+1}.$$

But in the last row we have $a_i = a_0 \sqcap a_1$. Thus the last row gives

$a_0x_i + a_1y_i = a_0 \sqcap a_1$.

By using 2.12 we can solve linear equations of the form

3) $ax + by = c$, $\quad$ a, b, c integers, with x and y integers. **First** we note that any divisor of a and b is also a divisor of c, and may be divided out. i.e. we may assume in 3) that

4) $a \sqcap b = 1$.

Then the equation

5) $ax + by = 1$

is soluble in integers. The algorithm 2.12 gives the solution (x_0, y_0). From this all solutions of 5) may easily be found. Let (x, y) be another solution.

Then $ax + by = 1$

$ax_0 + by_0 = 1$

Subtracting and rearranging we have

$$6)\quad \frac{x - x_0}{y - y_0} = \frac{-b}{a}$$

By 4), there exists an integer t, such that

$x - x_0 = -bt, \quad y - y_0 = at$

or

$$7)\quad \left.\begin{aligned} x &= x_0 - bt \\ y &= y_0 + at \end{aligned}\right\} \quad t \in \mathbb{Z}$$

In 7) we exhibit all solutions of 5). Then all the solutions of 3) are

$$8)\quad \left.\begin{aligned} x &= c(x_0 - bt) \\ y &= c(y_0 + at) \end{aligned}\right\} \quad t \in \mathbb{Z}$$

Exercises

1. Write a BASIC program for Fig. 2.12 and determine in 1) the values

   d, x, y for

   a) a = 91, b = 56 b) a = 286, b = 121 c) a = 119, b = 13

   d) a = 144, b = 89 e) a = 233, b = 144 f) a = 377, b = 233

   g) a = 610, b = 377 h) a = 927, b = 610

   In d) to h) occur consecutive members of the Fibonacci sequence. What formula may be conjectured from the values printed out?

2. The cheque problem. Prof. X changes a cheque for £x and y pence. The cashier pays him as though it read £y and x pence; as a result Prof. X receives 5 pence more than twice the correct amount. Determine x and y.

2.4. Prime numbers

A natural number is a prime number if it has exactly two divisors. Euclid (365?- 300? BC) knew that there are infinitely many prime numbers. The sequence of primes 2, 3, 5, 7, 11, 13, 17, 19, 23, 29,,, is very irregular. On the one hand there are arbitrarily large gaps between primes. For example, $n! + 2$, $n! + 3$, ... , $n! + n$ are $n - 1$ consecutive composite numbers . On the other hand, it is very probable that there are infinitely many twin-primes i.e. prime number pairs of the form $(p, p + 2)$.

Every prime > 3 has the form $6n - 1$ or $6n + 1$, since for $n \geq 1$, numbers of the form $6n$, $6n + 2$, $6n + 3$ are clearly not prime. Thus all prime pairs other than $(3, 5)$ have the form $(6n - 1, 6n + 1)$.

If $n > 1$ is not a prime it can be decomposed into non-trivial factors.

$n = d_1 d_2$, $1 < d_1 < n$, $1 < d_2 < n$.

The divisors of d_1 and d_2 cannot both be $> \sqrt{n}$, otherwise $d_1 d_2 > n$. A composite number n thus has a non-trivial divisor $d \leq \sqrt{n}$. In other words: If n has no divisors d in $1 < d \leq \sqrt{n}$, then n is a prime.

We wish to make a table of primes. All relevant methods depend upon division or sieving.

A table of primes by division

We wish to print all primes $\leq$ N. The primes 2 and 3 play a special role and are best printed out separately. Let A be a number under test and B a possible divisor. Initially A = 5 and B = 3. We need to divide A by all the primes $\leq \sqrt{A}$. If any division gives a whole number, then A is not prime. Otherwise A is a prime and is printed. Afterwards $A \leftarrow A + 2$ and $B \leftarrow 3$ are performed and the cycle is repeated. The test divisor B need only run through the prime numbers $\leq \sqrt{A}$. Since it is inconvenient to store all the primes $\leq \sqrt{A}$ we allow B to run through all the odd numbers $\leq \sqrt{A}$. Fig. 2.13 gives the flow diagram and the BASIC program for N = 1000.

```
10 PRINT 2;3;
20 FOR A = 5 TO 1000 STEP 2
30        FOR B = 3 TO SQR (A) STEP 2
40              IF A/B = INT (A/B) THEN 70
50        NEXT B
60        PRINT A;
70 NEXT A
80 END
```

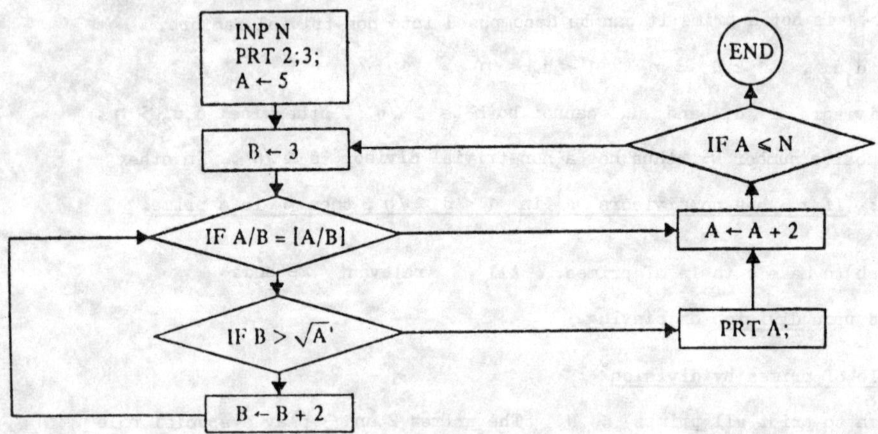

2 3 5 7 11 13 17 19 23 29 31 37 41 43 47 53 59 61 67 71 73 79 83 89 97 101
103 107 109 113 127 131 137 139 149 151 157 163 167 173 179 181 191 193 197
199 211 223 227 229 233 239 241 251 257 263 269 271 277 281 283 293 307 311
313 317 331 337 347 349 353 359 367 373 379 383 389 397 401 409 419 421 431
433 439 443 449 457 461 463 467 479 487 491 499 503 509 521 523 541 547 557
563 569 571 577 587 593 599 601 607 613 617 619 631 641 643 647 653 659 661
673 677 683 691 701 709 719 727 733 739 743 751 757 761 769 773 787 797 809
811 821 823 827 829 839 853 857 859 863 877 881 883 887 907 911 919 929 937
941 947 953 967 971 977 983 991 997

Fig. 2.13

Exercises

1. In each execution of line 40 the left and right sides of the equality are both divided. It is necessary to divide only once if line 40 is replaced by the two lines

```
40   C = A/B
45   IF  C = INT(C) THEN 70
```

Compare the computation times of the old and new programmes for N = 2000. Remove the print instructions from both programs beforehand.

2. Write a program to count the number of primes in the interval from M to N ($M > 3$ and odd). How many primes are there between 1000 and 2000, 2000 and 3000, 3000 and 4000, 10000 and 11000?

3. Write a program which prints the gaps between successive primes up to 1000. i.e. 1, 2, 2, 4, 2, 4, 2, 4, 6, 2, 6, ...

4. Write a program to print the largest gap between consecutive primes in the interval between M and N ($M > 2$). Suppose the smallest prime P in the interval (M, N) is known.

5. Write a program which locates the first gap of 16 between successive primes. i.e. it gives the smallest prime pair of the form (p, p + 16).

Twin Primes

We write a fast program which determines all prime pairs between A and B. All twins other than (3, 5) have the form ($6n - 1$, $6n + 1$). For the first pair (X, X + 2) having the form ($6n - 1$, $6n + 1$) in the interval from A to B we have $X = 6\left[\frac{A}{6}\right] + 5$. Y is a test divisor. Initially Y = 5, since $6n \pm 1$ is not divisible by 3. Y runs through the odd numbers from 5 to $\sqrt{X} + 1$. The smaller number $\sqrt{X + 2}$ would suffice but by using $\sqrt{X} + 1$ we guard against round-off errors. The decisive idea is to test the elements of the pair (X, X + 2) at the same time. This occurs in line 40 (Fig. 2.14). Usually one element of the pair has a small factor. Then the test can be abandoned and the value of X increased by 6.

```
10 INPUT A, B
20 FOR X = 6*INT (A/6) + 5 TO B STEP 6
30      FOR Y = 5 TO SQR (X) + 1 STEP 2
40           IF X = Y*INT (X/Y) OR X + 2 = Y*INT ((X + 2)/Y) THEN 70
50      NEXT Y
60      PRINT X; X + 2,
70 NEXT X
80 END
```

1019 1021 1031 1033 1049 1051 1061 1063 1091 1093

10007 10009 10037 10039 10067 10069 10091 10093

100151 100153 100361 100363 100391 100393 100517 100519
100547 100549 100799 100801

1000037 1000039 1000211 1000213 1000289 1000291
1000427 1000429 1000577 1000579 1000619 1000621
1000667 1000669 1000721 1000723 1000847 1000849
1000859 1000861 1000919 1000921

Fig. 2.14

The program in Fig. 2.14 was run for the intervals

$(10^3, 10^3 + 100), (10^4, 10^4 + 100), (10^5, 10^5 + 1000), (10^6, 10^6 + 1000)$.

There are surprisingly many twin primes. Their density certainly declines slowly but in an irregular way. Between 10^5 and $10^5 + 100$ there are only 6 pairs, while between 10^6 and $10^6 + 100$ there are eleven pairs. Further information on the distribution of twin primes can be found in [5].

Decomposition into factors

Every natural number $n > 1$ may be expressed uniquely in the form

$n = p_1 p_2 p_3 \cdots p_m$, $p_1 \leq p_2 \leq p_3 \cdots \leq p_m$ where each p_i is a prime.

We write a program which finds this prime decomposition. We first describe the method in English. We divide n in turn by p = 2, 3, 5, 7... If the division has no remainder we print p, set $n \leftarrow \frac{n}{p}$ and attempt to divide the new n by p and larger primes. As soon as $\left[\frac{n}{p}\right] < p$ we can stop since n is prime.

It is easier to use as divisors not just the primes but instead 2, 3 and all numbers of the form $6n \pm 1$. i.e. beginning with D = 5 we take alternately 2 and 4 steps. In the program the changing step of 2 or 4 in the divisor is controlled by the switch S. The factors 2 and 3 must be completely removed first. This plan is realised in Fig. 2.15.

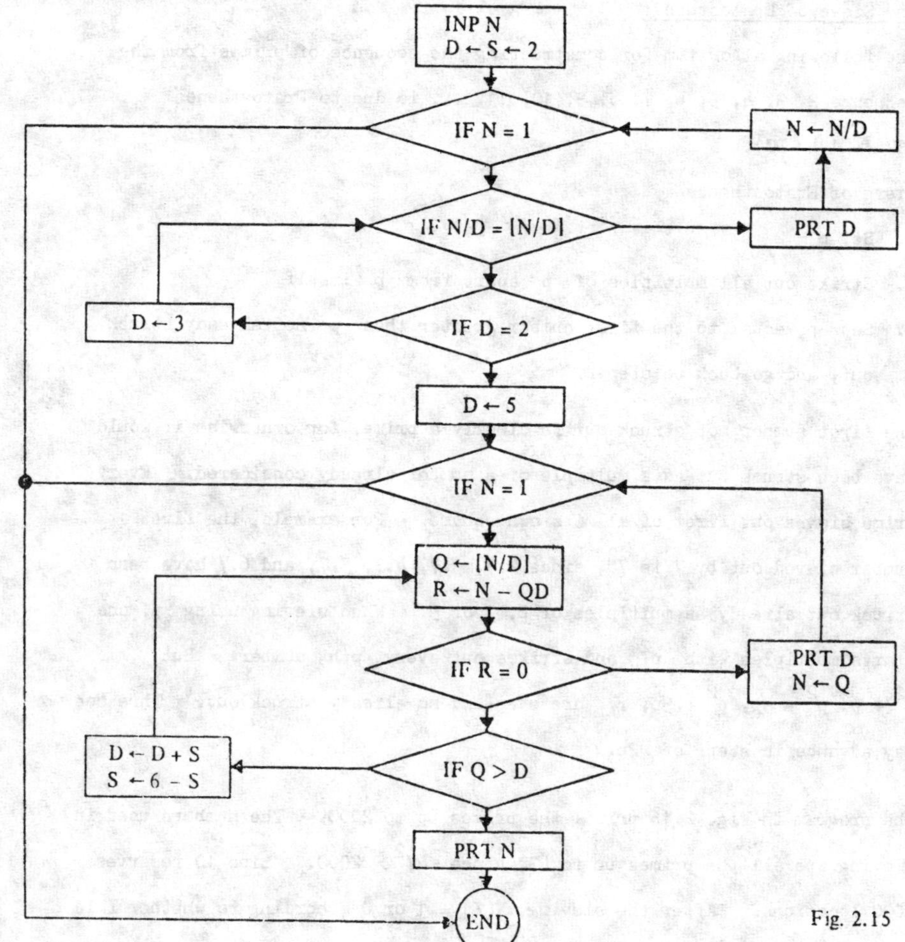

Fig. 2.15

Exercises

6. Convert Fig. 2.15 into BASIC and decompose the numbers 123456789, 987654321, $2^{32}+1$, 1264460, 81128632, 600000017.

7. We wish radically to simplify the program of Fig. 2.15 by allowing the test divisor D to run from 2 to $\sqrt{n}$ each time. Write the corresponding BASIC program and compare the computation times for the decomposition of 600000017.

8. The numbers 1979339333 and 1979339339 are primes. Show that they remain primes if digits are removed one at a time from the right hand end.

The Sieve of Eratosthenes

The following algorithm for constructing the sequence of primes from the sequence 2, 3, 4, 5, 6, 7, 8, 9, 10, 11, ... is due to Eratosthenes (276?- 194? BC).

Sieve of Eratosthenes

1. Set p = 2
2. Strike out all multiples of p apart from p itself
3. Set p equal to the first number greater than p not already struck out, and go back to step 2.

The first number not struck out is clearly a prime, for otherwise it would have been struck out as a multiple of a number already considered. Every prime sieves out first of all its own square. For example, the first number sieved out by 7 is 7^2, since 2.7, 3.7, 4.7, 5.7, and 6.7 have been struck out already as multiples of 2, 3 or 5. When sieving using p, one starts multiples with p^2 and strikes out every pth number. But $p^2 + p$, $p^2 + 3p$, $p^2 + 5p$, ... are even and so already struck out. Thus one may advance in steps of 2p.

The program in Fig. 2.16 prints the primes up to 2000. The numbers used in sieving are all the primes up to 43, since $47^2 > 2000$. Line 10 reserves 2000 locations. After the sieving, X(I) = 1 or 0 according to whether I is a prime number or not. The single even prime number 2 is printed separately. Line 20 places 1 in the locations with odd numbers. These are our candidates as primes. 30 reads the first unused sieve-number p from line 90. Row 40 deletes p^2, $p^2 + 2p$, $p^2 + 4p$, ..., 50 reads the next prime number from line 90, or else prints 2 when the data is exhausted. 60 - 80 print all the numbers not struck out, up to 2000.

```
 10 DIM X (2000)
 20 FOR I = 3 TO 2000 STEP 2 X (I) = 1
 30 READ P
 40 FOR I = P*P TO 2000 STEP 2*P X (I) = 0
 50 IF P < 43 THEN 30 ELSE PRINT 2;
 60 FOR I = 3 TO 2000 STEP 2
 70      IF X (I) <> 0 THEN PRINT I;
 80 NEXT I
 90 DATA 3, 5, 7, 11, 13, 17, 19, 23, 29, 31, 37, 41, 43
100 END
```

```
2 3 5 7 11 13 17 19 23 29 31 37 41 43 47 53 59 61 67 71 73 79 83 89 97 101
103 107 109 113 127 131 137 139 149 151 157 163 167 173 179 181 191 193
197 199 211 223 227 229 233 239 241 251 257 263 269 271 277 281 283 293 307
311 313 317 331 337 347 349 353 359 367 373 379 383 389 397 401 409 419 421
431 433 439 443 449 457 461 463 467 479 489 491 499 503 509 521 523 541 547
557 563 569 571 577 587 593 599 601 607 613 617 619 631 641 643 647 653 659
661 673 677 683 691 701 709 719 727 733 739 743 751 757 761 769 773 787 797
809 811 821 823 827 829 839 853 857 859 863 877 881 883 887 907 911 919 929
937 941 947 953 967 971 977 983 991 997 1009 1013 1019 1021 1031 1033 1039
1049 1051 1061 1063 1069 1087 1091 1093 1097 1103 1109 1117 1123 1129 1151
1153 1163 1171 1181 1187 1193 1201 1213 1217 1223 1229 1231 1237 1249 1259
1277 1279 1283 1289 1291 1297 1301 1303 1307 1319 1321 1327 1361 1367 1373
1381 1399 1409 1423 1427 1429 1433 1439 1447 1451 1453 1459 1471 1481 1483
1487 1489 1493 1499 1511 1523 1531 1543 1549 1553 1559 1567 1571 1579 1583
1597 1601 1607 1609 1613 1619 1621 1627 1637 1657 1663 1667 1669 1693 1697
1699 1709 1721 1723 1733 1741 1747 1753 1759 1777 1783 1787 1789 1801 1811
1823 1831 1847 1861 1867 1871 1873 1877 1879 1889 1901 1907 1913 1931 1933
1949 1951 1973 1979 1987 1993 1997 1999
```

Fig. 2.16

The program in Fig. 2.16 is wasteful since only the locations with odd numbers are used. Also it is worth noting that on some small calculators one cannot reserve so many locations. Often one can reserve only 256. So we re-write our program so that X(I) = 1 or 0 according to whether 2I + 1 is a prime or not. Then we can compile a table of primes up to 513. The numbers used in sieving are the primes up to 19, since $23^2 = 529$. You can check the correctness of the program in Fig. 2.17. Only line 60 introduces some new ideas.

```
10 DIM X (255)
20 FOR I = 1 TO 255
30       X (I) = 1
40 NEXT I
50 READ P
60 FOR I = (P*P – 1)/2 TO 255 STEP P
70       X (I) = 0
80 NEXT I
90 IF P< 19 THEN 50
100 PRINT 2;
110 FOR I = 1 TO 255
120       IF X (I) = 0 THEN 140
130       PRINT 2*I + 1;
140 NEXT I
150 DATA 3, 5, 7, 11, 13, 17, 19
160 END
```

```
2 3 5 7 11 13 17 19 23 29 31 37 41 43 47 53 59 61 67 71 73 79 83 89 97 101
103 107 109 113 127 131 137 139 149 151 157 163 167 173 179 181 191 193 197
199 211 223 227 229 233 239 241 251 257 263 269 271 277 281 283 293 307 311
313 317 331 337 347 349 353 359 367 373 379 383 389 397 401 409 419 421 431
433 439 443 449 457 461 463 467 479 487 491 499 503 509
```

Fig. 2.17

Fig. 2.17 is written in a simple version of BASIC which should be accepted by any small computer, whereas this is not true of Fig. 2.16.

Exercises

9. A natural number is called square-free if it is not divisible by the square of a prime number. Let $h(n)$ be the number, and $q(n) = \frac{h(n)}{n}$ be the proportion, of square-free numbers $\leq n$.

   a) Determine $h(n)$ and $q(n)$ by hand for $n = 10, 20, 30, 40, 50, 60$.

   b) Place the digit 1 in X(1) to X(2000). Sieve out all multiples of $2^2, 3^2, 5^2, 7^2, 11^2, \ldots 41^2, 43^2$ and print a table with the entries n, $h(n)$, $q(n)$ for $n = 100$ to 2000 in steps of 100.

   It can be shown that $\lim_{n \to \infty} q(n) = \frac{6}{\pi^2}$. See [5].

2.5. Periodic decimals

Gauss while a schoolboy found $\frac{1}{n}$ as a decimal for all $n < 1000$, in order to study the dependence of the period length $p(n)$ upon n.

Let n be co-prime with 10, i.e. $10 \sqcap n = 1$. We want to determine the period length of the decimal representation of $\frac{1}{n}$. Let

$$\frac{1}{n} = \frac{r_1}{n} = 0, a_1 a_2 a_3 a_4 \ldots ,$$

$r_1 = 1$ we call the first remainder. Multiplication by 10 gives

$$\frac{10r_1}{n} = a_1 . a_2 a_3 a_4 \ldots$$

$$a_1 = \left[\frac{10r_1}{n}\right], \quad \frac{10r_1}{n} - \left[\frac{10r_1}{n}\right] = 0.a_2 a_3 a_4 .$$

$$\frac{r_2}{n} = \frac{10r_1 - n . \left[\frac{10r_1}{n}\right]}{n} = 0.a_2 a_3 a_4$$

The second remainder is

$$r_2 = 10r_1 - n \left[\frac{10r_1}{n}\right] = 10r_1 \mod n.$$

i.e. the next remainder is obtained from the preceding one by the instruction

$$r \leftarrow 10r - n \left[\frac{10r_1}{n}\right] \quad \text{or} \quad r \leftarrow 10r \mod n$$

Initially $r = 1$. Therefore the sequence of remainders is given by

$1, 10, 10^2, 10^3, \ldots \pmod n$.

Since only the remainders $1, 2, \ldots, n - 1$ are possible, this sequence is periodic with period length $\leq n - 1$. Repetition begins at the earliest moment i.e. the remainder 1 is the first to be repeated. In fact, suppose $10^i \equiv 10^k \pmod n$, $i < k$.

Since $10 \sqcap n = 1$, the congruence can be divided by 10, and so

$$10^{i-1} \equiv 10^{k-1}, \quad 10^{i-2} \equiv 10^{k-2}, \ \ldots \ 1 \equiv 10^{k-i} \pmod n.$$

The period length is therefore the smallest index p such that

$10^p \equiv 1 \pmod n$.

The program in Fig. 2.18 prints the period length p for every n, $n < 100$, $n \sqcap 10 = 1$.

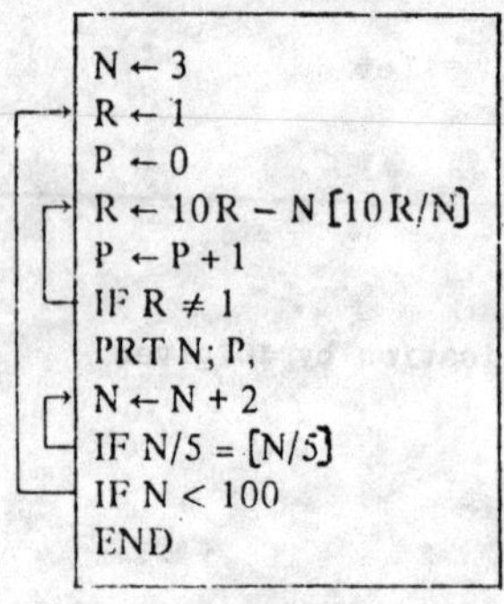

Fig. 2.18

Exercises

1. Write a BASIC program for Fig. 2.18 and examine the table produced by the computer.

   a) What is the connection between n and p(n) when n is a prime?

   [b)] Try to find a connection between n and p(n) for composite numbers.

2. Write a program analogous to Fig. 2.18 for base 2. Compare the period lengths with those in base 10.

   Hint $\frac{1}{n}$ is pure periodic in base 2 if $2 \sqcap n = 1$. The period length is the smallest index p such that

   $2^p \equiv 1 \pmod{n}$.

3. Write a program which prints the period (recurring part) of $\frac{1}{n}$ $(n \sqcap 10 = 1)$.

2.6. Continued fractions

Let x be a real number. If x is not an integer we can write

$$x = z_0 + \frac{1}{x}, \quad z_0 = [x], \quad x_1 = \frac{1}{x - z_0} > 1.$$

If x_1 is not integral, we apply the same transformation to x_1

$$x_1 = z_1 + \frac{1}{x_2}, \quad z_1 = [x_1], \quad x_2 = \frac{1}{x_1 - z_1} > 1.$$

For rational values of x we obtain an expression of the form

$$x = z_0 + \cfrac{1}{z_1 + \cfrac{1}{z_2 + \cfrac{1}{z_3 + \cfrac{1}{\ddots + \cfrac{1}{z_n}}}}} = z_0 + 1/(z_1 + 1/(z_2 + 1/(z_3 + \ldots + 1z_n \ldots)))$$

This expression is called a continued fraction. It is also denoted by $[z_0 ; z_1, z_2, \ldots , z_n]$. The z_i are called partial quotients and $r_i = [z_0 ; z_1, z_2 \ldots z_i]$ is the ith convergent. For irrational x, the continued fraction does not terminate. Continued fractions are easily evaluated (beginning at the tail end) using a pocket calculator. The program in Fig. 2.19 prints the sequence z_i when x is input. For $x = \pi$, $\sqrt{2}$, $(\sqrt{5}+1)/2$, e it gives the sequences shown. Because of rounding errors the underlined quotients are incorrect

```
INP X
PRT [X]
IF X = [X]
X ← 1/(X − [X])
END
```

$$\pi = [3;7,15,1,292,1,1,1,\underline{4,1,2,14,1,2,2,1,3},\ldots]$$

$$\sqrt{2} = [1;2,2,2,2,2,2,2,2,2,2,2,2,2,2,\underline{3,1,3,2,2},\ldots]$$

$$(\sqrt{5}+1)/2 = [1;1,1,1,1,1,1,1,1,1,1,1,1,1,1,1,1,1,1,1,1,1,1,1,1,1,1,1,\underline{7,1,2,4,3},\ldots]$$

$$e = [2;1,2,1,1,4,1,1,6,1,1,8,1,1,\underline{11,1,1,2,2,1,1,1,1,1},\ldots]$$

Fig. 2.19

The convergents for $\pi = 3.141592653589793\ldots$ are

$$r_0 = 3, \quad r_1 = 3\tfrac{1}{7}, \quad r_2 = 3\,\frac{15}{106}, \quad r_3 = \frac{355}{113} = 3\,\frac{16}{113}$$

These are well-known approximations for π. The pocket calculator gives

$r_0 = 3.000000000$, $r_1 = 3.142857143$, $r_2 = 3.141509434$

$r_3 = 3.141592920$, $r_4 = 3.141592653$, $r_5 = 3.141592654$

1. Example

Connection with the Euclidean algorithm

We use the Euclidean algorithm with a = 67 and b = 29:

$$\left.\begin{aligned} 67 &= 29.\underline{2} + 9 \Rightarrow \frac{67}{29} = 2 + \frac{9}{29} = 2 + \cfrac{1}{\frac{29}{9}} \\ 29 &= 9.\underline{3} + 2 \Rightarrow \frac{29}{9} = 3 + \frac{2}{9} = 3 + \cfrac{1}{\frac{9}{2}} \\ 9 &= 2.\underline{4} + 1 \Rightarrow \frac{9}{2} = 4 + \frac{1}{2} \end{aligned}\right\} \Rightarrow \frac{67}{29} = 2 + \cfrac{1}{3 + \cfrac{1}{4 + \frac{1}{2}}} = [2 ; 3, 4, 2]$$

$$2 = 1.\underline{2} + 0$$

The quotients appearing in the Euclidean algorithm are also those of the continued fraction expression for $\frac{a}{b}$.

2. Example :

The continued fraction expression for $\log_b a$

Let $a > 1$ and $b > 1$. We wish to calculate $x > 0$ in

1) $b^x = a$

If we write $x = z_0 + \frac{1}{x_1}$, $z_0 = [x]$, $x_1 > 1$ then $b^{z_0 + \frac{1}{x_1}} = a$.

If $a \leftarrow \frac{a}{b}$ is performed z_0 times, the resulting equation is

$b^{\frac{1}{x_1}} = a$ or $b = a^{x_1}$.

If we interchange the values of a and b, we have the original problem 1). This leads to the following algorithm, which prints the sequence $z_0, z_1, z_2, \ldots$ of the continued fraction expression.

1) **Set** $z \leftarrow 0$
2) **While** $a \geqslant b$, repeat $a \leftarrow \frac{a}{b}$; $z \leftarrow z + 1$
3) **Print** z and exchange a and b.
4) **If** $b > 1$ go to step 1, otherwise STOP

For $b = 10$, $a = 2$ we have $\lg 2 = [0; 3, 3, 9, 2, 2, 4, 6, 2, 1, 1, 3, \ldots.]$
The pocket calculator gives $\lg 2 = 0.3010299957$, in which all the digits are correct. Fig. 2.20 shows the corresponding program. It produces some errors in the final quotients. These influence only the 11th and higher places of decimals.

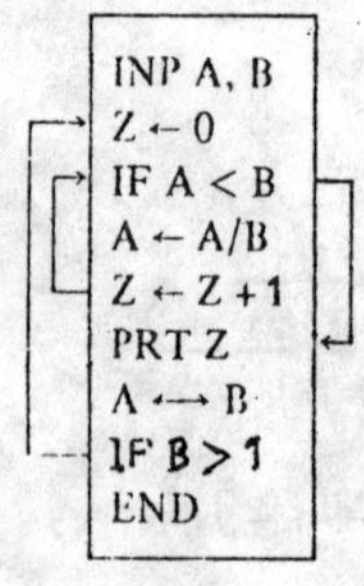

INP A, B
Z ← 0
IF A < B
A ← A/B
Z ← Z + 1
PRT Z
A ←→ B
IF B > 1
END

Fig. 2.20

Exercises

1 a) Let a, b be natural numbers. Write a program which when a, b are input, prints the quotients of the continued fraction expression for $\frac{a}{b}$ using the Euclidean algorithm.

b) Test the program for consecutive Fibonacci numbers in Fig. 1.62, i.e. for $a = F_{n+1}$, $b = F_n$. What happens? Proof!

2. 1000 random numbers are chosen from (0, 1). For each number X, $Z = [1/X]$, the first quotient of its continued fraction expression, is determined. How often do Z = 1, 2, 3,... 10 occur? Let p_n be the probability that Z = n. Try to discover a formula for p_n.

2.7* Chinese Prime numbers

If a is a divisor of b we write $a | b$ (read 'a divides b'). Otherwise we write $a \nmid b$ (read 'a does not divide b'). 2500 years ago in China the following conjecture was put forward:

1) n is a prime number $\iff n | 2^n - 2$, $n > 1$. We shall call $n (> 1)$ a Chinese prime number (pseudoprime) if $n | (2^n - 2)$. Fermat showed in 1640 that

2) n is a prime number $\Rightarrow n | (a^n - a)$ for all integers a.

The converse is unfortunately untrue. If it were correct, we would have a very quick test for primeness. The computation time for testing the relation $n | (a^n - a)$ is proportional to $\log_2 n$. Against that, the prime number test by division using 2, 3, 5, ... $[\sqrt{n}]$ requires a time proportional to $\sqrt{n}$.

In order to test the relation $n | (a^n - a)$, one calculates $a^n - a$ modulo n, and for this the fast powering program (Fig. 2.5b) is used.

Exercises

1. Write a program which prints the Chinese prime numbers upto 1000.
2. Write a program which prints all composite Chinese prime numbers under

1000, that is, counter examples for the conjecture 1).

3. The statement

2a) $n \nmid (a^n - a) \Rightarrow n$ is not prime

is equivalent to 2).

In 1640 Fermat in a letter to Mersenne suggested the conjecture $2^{32} + 1$ is prime. Show, using 2a) with $a = 3$, $n = 2^{32} + 1$, that n is not a prime. (A computer is needed which can calculate to 20 decimal places).

<u>Remark</u>: Using 2a) one can often show quickly that a number n is not prime.

4. a) In Anchuria 2^{11} years ago it was conjectured that

3) n is prime $\Leftrightarrow n \mid (3^n - 3)$

b) In Sikinia 3^7 years ago it was conjectured that

4) n is prime $\Leftrightarrow n \mid (5^n - 5)$

Find all the counter examples less than 1000 for 3) and 4).

<u>Remark</u>: There are composite numbers n such that $n \mid (a^n - a)$ for all integers a. Find one such <u>absolute pseudoprime</u> which disproves the conjectures of all nationalities.

3. GEOMETRY

In this chapter we consider geometric problems. Central among these are the calculation of π and algorithms for the functions defined in terms of the circle $x^2 + y^2 = 1$ and the hyperbola $xy = 1$.

3.1. Quadrature of the parabola by the method of Archimedes

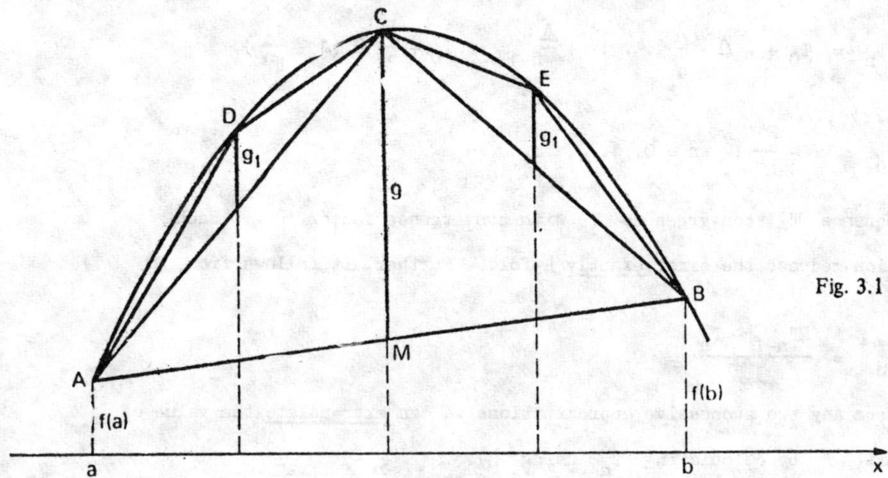

Fig. 3.1

Let the region under the parabola $f(x) = px^2 + q + r$ from a to b have area T. (Fig. 3.1). We construct a sequence $T_0, T_1, T_2, \ldots$ of approximations for T by approximating the region using 1, 2, 4, 8, ... 2^n ... trapezia of equal width. We have $T_0 = h\,\dfrac{f(a) + f(b)}{2}$, $h = b - a$.

The first error $T - T_0$ is equal to the area P of the segment of the parabola bounded by AB. A simple calculation shows that

1) $\overline{CM} = g = f\left(\dfrac{a + b}{2}\right) - \dfrac{f(a) + f(b)}{2} = \dfrac{-p}{4}(b - a)^2 = \dfrac{-p}{4}h^2.$

ΔABC has area $\Delta = \dfrac{gh}{2}$. Further $T_1 = T_0 + \Delta$. From 1),

$g_1 = \dfrac{-p}{4} \cdot \left(\dfrac{h}{2}\right)^2 = \dfrac{g}{4}.$

Thus ΔADC and ΔBEC together have area $\dfrac{\Delta}{4}$ and $T_2 = T_1 + \dfrac{\Delta}{4}.$

By analogy we obtain

$$T_{n+1} = T_n + \frac{\Delta}{4^n}, \quad n = 0, 1, 2, \ldots$$

The parabolic segment has area

$$2) \quad P = \Delta + \frac{\Delta}{4} + \frac{\Delta}{4^2} + \ldots = \frac{4}{3}\Delta = \frac{2}{3} gh.$$

Thus $T = T_0 + \frac{4}{3}\Delta$

$$T_n = T_0 + \Delta + \frac{\Delta}{4} \ldots + \frac{\Delta}{4^{n-1}} = T_0 + \frac{4}{3}\Delta\left(1 - \frac{1}{4^n}\right).$$

Or

$$3) \quad T_n = T - \frac{P}{4^n}, \quad n = 0, 1, 2, \ldots$$

The sequence T_n converges to T with convergence factor $\frac{1}{4}$. Each iteration reduces the error <u>exactly</u> 4-fold. Further, it follows from 3) that

$$4) \quad T_n' = \frac{4T_{n-1} - T_n}{3} = T.$$

Thus from any two successive approximations we can <u>extrapolate</u> the value of the limit T by calculating T_n' using 4).

A short arc of a curve will differ very little from a parabolic arc. If we replace the parabola in Fig. 3.1 by another curve and construct the trapezium sequence T_n, the errors $T - T_n$ will be approximately 4 times smaller with each iteration. The value of T_n' obtained from 4) will usually not give T <u>exactly</u>, but it will lie much closer to T than does T_n.

3.2. Calculation of π by Archimedes

The number π is one of the most important and famous numbers. There is no method of calculating it which is both simple and elementary. The struggle involving π has a history spanning 4000 years. The Old Testatment uses $\pi = 3$ (I Kings 7:23, II Chronicles 4:2). The Babylonians used $\pi = 3$ and $\pi = 3\frac{1}{8}$. The Egyptian writer Ahmes (1700 BC?) gives the following algorithm for calculating the area of a circle:

Subtract from the diameter of the circle $\frac{1}{9}$ of the diameter and square the result.

Exercise 1 :

What approximation did the Egyptians use for π ?

A circle with radius 1 has area π and half the circumference is π.

The elementary methods of determining π calculate successive approximations to the circumference or the area of a unit circle. Among the first to calculate an approximation for π was Archimedes (about 260 B.C.). He approximated one half of the circumference of a unit circle using inscribed and circumscribed regular polygons. We wish to determine the length L of the arc AB in Fig. 3.2., using equal chords and tangents respectively to approximate AB by interior and exterior polygonal arcs of 1, 2, 4, ...,2^n sides.

Let the length of a chord-segment and of a tangent-segment be $2s_n$ and $2t_n$ respectively, and let s_n be distant c_n from the centre of the circle. Let the total lengths of the interior and exterior polygonal arcs be S_n and T_n respectively. Then we have

(1) $S_n = 2^{n+1}s_n$, $T_n = 2^{n+1}t_n$, $S_n < L < T_n$, $\lim_{n\to\infty} S_n = \lim_{n\to\infty} T_n = L$.

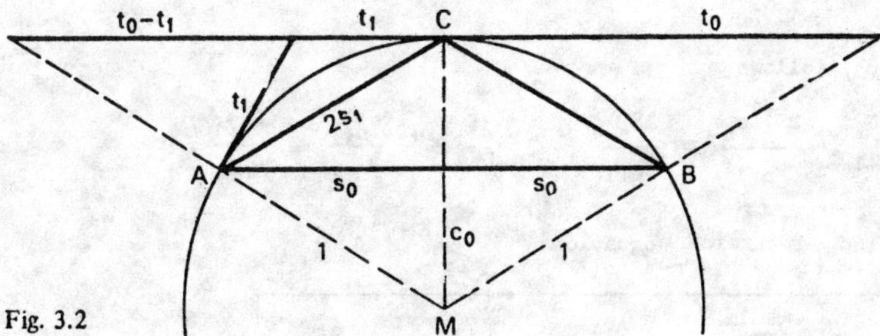

Fig. 3.2

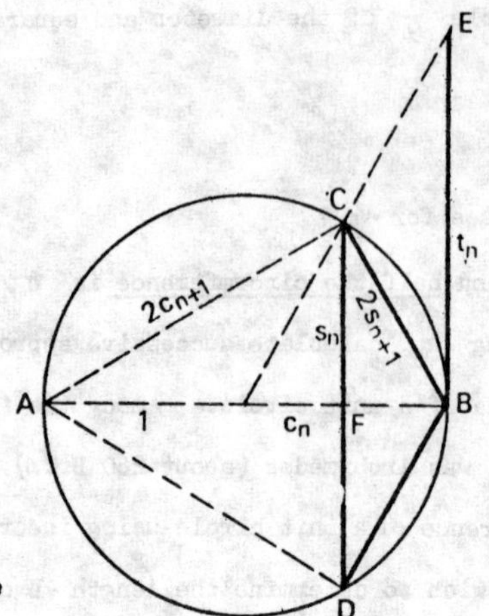

Fig. 3.3

In Fig. 3.3, the area of the kite ADBC may be determined in two ways, from which we obtain $2s_n = 2s_{n+1} \cdot 2c_{n+1}$, or

(2) $s_{n+1} = \dfrac{s_n}{2c_{n+1}}$

Further, $t_n/1 = s_n/c_n$ or

(3) $t_n = \dfrac{s_n}{c_n}$.

By similar triangles, $\overline{AC}^2 = \overline{AB} \cdot \overline{AF}$, i.e. $4c_{n+1}^2 = 2(1+c_n)$, or

(4) $c_{n+1} = \sqrt{\dfrac{1+c_n}{2}}$.

From (1) and (3) follows

$$S_{n+1} = 2^{n+2}s_{n+1} = \frac{2^{n+1}s_n}{c_{n+1}} = \frac{S_n}{c_{n+1}} , \quad T_{n+1} = 2^{n+2}\frac{s_{n+1}}{c_{n+1}} = \frac{S_{n+1}}{c_{n+1}} .$$

Thus we have the recursion equations

(5) $$\boxed{c_{n+1} = \sqrt{\frac{1+c_n}{2}} , \quad S_{n+1} = \frac{S_n}{c_{n+1}} , \quad T_{n+1} = \frac{S_{n+1}}{c_{n+1}}}$$

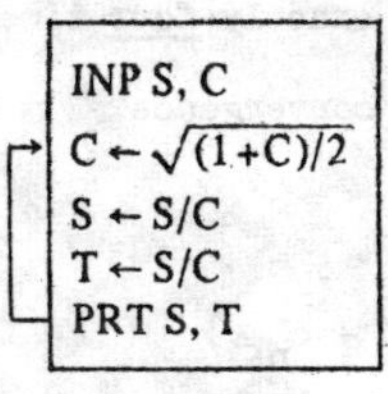

Fig. 3.4

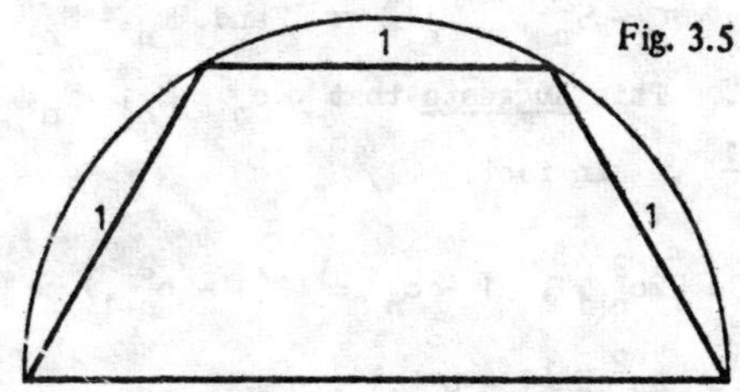

Fig. 3.5

They lead to the program in Fig. 3.4. The length S of the chord AB and its distance C from the centre are input. The program prints an <u>interval approximation</u> for the arc-length L. In order to have $L = \pi$, we must input S = 2, C = 0. Archimedes began as in Fig. 3.5, i.e. with $S = 3$, $C = \frac{\sqrt{3}}{2}$, and he carried out four iterations.

```
10 INPUT S, C
20 C = SQR ((1 + C)/2)
30 S = S/C
40 T = S/C
50 PRINT S, T, 1 – C, π – S, T – π
60 IF S < T THEN 20
70 END
```

S	T	1 – C	π – S	T – π
3.105828541	3.215390309	.034074174	.035764112	.073797656
3.132628613	3.159659942	.008555139	.008964040	.018067289
3.139350203	3.146086215	.002141077	.002242451	.004493562
3.141031951	3.142714600	.000535413	.000560703	.001121946
3.141452472	3.141873050	.000133862	.000140181	.000280396
3.141557608	3.141662747	.000033466	.000035046	.000070093
3.141583892	3.141610177	.000008367	.000008761	.000017523
3.141590463	3.141597034	.000002092	.000002190	.000004381
3.141592106	3.141593749	.000000523	.000000548	.000001095
3.141592517	3.141592927	.000000131	.000000137	.000000274
3.141592619	3.141582722	.000000033	.000000034	.000000068
3.141592645	3.141592671	.000000008	.000000009	.000000017
3.141592651	3.141592658	.000000002	.000000002	.000000004
3.141592653	3.141592654	.000000001	.000000001	.000000001
3.141592654	3.141592654	.000000000	.000000000	.000000000

Fig. 3.6

Now we need a suitable criterion for stopping. It is always the case that $S < T$. In Fig. 3.6 iteration was stopped as soon as this inequality failed through rounding errors. It is noticeable that with each iteration,

$1 - c_n$, $\pi - S_n$, $T_n - \pi$ and $T_n - S_n$ become almost exactly <u>four</u> times smaller. This <u>suggests</u> that c_n, S_n, T_n, $T_n - S_n$ have convergence factor $\frac{1}{4}$. In fact

$$1 + c_n = 2c_{n+1}^2 , \quad 1 - c_n = 2(1 - c_{n+1}^2) = 2(1 - c_{n+1})(1 + c_{n+1})$$

$$= 4c_{n+2}^2 (1 - c_{n+1}) .$$

Since $\lim\limits_{n \to \infty} c_n = 1$, $T_n - S_n = T_n(1 - c_n)$, and $S_n - S_{n-1} = S_n(1 - c_n)$

it follows that as $n \to \infty$,

$$\frac{1 - c_{n+1}}{1 - c_n} = \frac{1}{4c_{n+2}^2} \to \frac{1}{4} , \quad \frac{T_{n+1} - S_{n+1}}{T_n - S_n} = \frac{c_n}{c_{n+1}^2} \cdot \frac{1-c_{n+1}}{1-c_n} \to \frac{1}{4} ,$$

and $$\frac{S_{n+1} - S_n}{S_n - S_{n-1}} = \frac{1}{c_{n+1}} \cdot \frac{1-c_{n+1}}{1-c_n} \to \frac{1}{4}$$

3.3 Algorithms for the trigonometric functions

a) With the help of Archimedes' method we can construct simple algorithms for the trigonometric functions.

Fig. 3.7 shows the definition of the functions $s(x) = \sin x$ and $c(x) = \cos x$. In the interval $\frac{-\pi}{2} \leq x \leq \frac{\pi}{2}$, s is increasing and therefore invertible. The inverse function $g = s^{-1}$ is called arc sin x [(†) $\sin^{-1} x$].

Fig. 3.8 shows the definition of $g(x) = \text{arc sin } x$. In the interval $0 \leq x \leq \pi$, c is decreasing and so invertible. The inverse function $h = c^{-1}$ is called arc cos x [(†) $\cos^{-1} x$].

Fig. 3.9 shows the definition of $h(x) = \text{arc cos } x$. Fig. 3.8 shows that arc sin x can be calculated using Fig. 3.4. If $S = 2x$, and $C = \sqrt{1-x^2}$ are input, the output is 2 arc sin x.

† (translator's note)

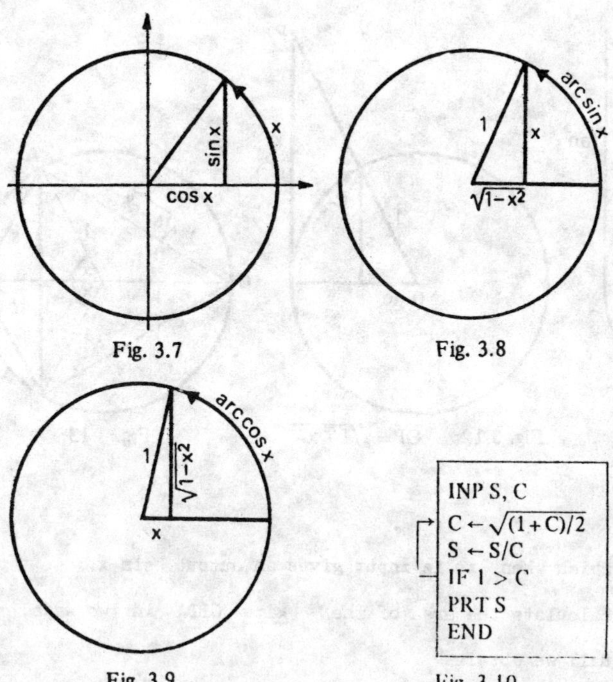

Fig. 3.7 Fig. 3.8

Fig. 3.9 Fig. 3.10

If instead $S = x$ is input, the output is arc sin x.

We simplify the program further, by omitting the variable T. The result is Fig. 3.10. For the input $S = x$, $C = \sqrt{1-x^2}$, this gives output arc sin x ; further, Fig. 3.9 shows that the input $S = \sqrt{1-x^2}$, $C = x$ gives output arc cos x.

(b) The Figures 3.11 and 3.12 give the definitions of tan x and arc tan x. In Fig. 3.12, $s = \dfrac{x}{\sqrt{1+x^2}}$ and $c = \dfrac{1}{\sqrt{1+x^2}}$.

Thus we can also calculate arc tan x using Fig. 3.10 , if we input

$$C = \frac{1}{\sqrt{1+x^2}} \quad \text{and} \quad S = \frac{x}{\sqrt{1+x^2}} .$$

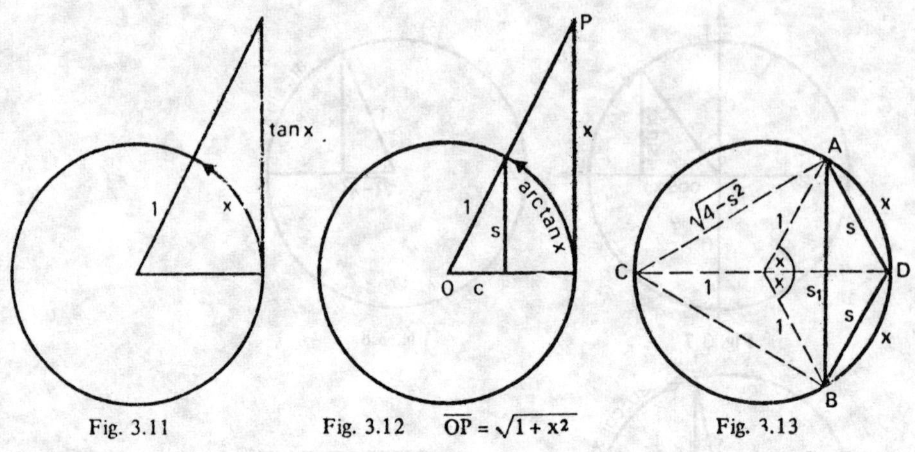

Fig. 3.11 Fig. 3.12 $\overline{OP} = \sqrt{1+x^2}$ Fig. 3.13

(c) We seek a program which when x is input gives as output $\sin x$. In Fig. 3.13, one can calculate the area of the <u>kite</u> CBDA in two ways. Setting the results equal, we obtain

$$(1) \quad s_1 = s\sqrt{4-s^2}$$

Thus through the assignment $s \leftarrow s\sqrt{4-s^2}$ one can calculate the sine of the double angle $2x$, subtended at the centre, if the sine of the angle x is known. We approximate the arc $\widehat{AB} = 2x$ by 2^n equal chords. If n is fairly large $(n > 16)$, then each of the chords has length given very accurately by $s = \frac{2x}{2^n}$. After n substitutions $s \leftarrow s\sqrt{4-s^2}$ we have $s = \widehat{AB} = 2\sin x$. This gives the program in Fig. 3.14.

If x is input in radians $\sin x$ is printed. If x is given in degrees, it may be first converted to radians by $x \leftarrow \pi \frac{x}{180}$.

The program in Fig. 3.14 was tested by calculating $\sin\frac{\pi}{6} = 0{\cdot}5$, $\sin\frac{\pi}{2} = 1$, and $\sin\frac{\pi}{180} = 0.01745240644$ for different input values of n. Fig. 3.15 shows the result

```
INP X, N
S ← 2X/2↑N
S ← S √(4 − SS)
N ← N − 1
IF N > 0
PRT S/2
END
```

Fig. 3.14

n \ x	$\pi/6$	$\pi/2$	$\pi/180$
1	0.5053368618	0.9723086202	0.01745262794
2	0.5013044765	0.9990556716	0.01745246181
3	0.5003243298	0.9999472391	0.01745242028
4	0.5000809716	0.9999967886	0.01745240990
5	0.5000202360	0.9999998006	0.01745240730
6	0.5000050586	0.9999999876	0.01745240665
7	0.5000012646	0.9999999992	0.01745240649
8	0.5000003162	1.0000000000	0.01745240645
9	0.5000000791		0.01745240644
10	0.5000000198		
11	0.5000000050		
12	0.5000000012		
13	0.5000000003		
14	0.5000000001		
15	0.5000000000		

Fig. 3.15

(d) We construct a sine program which is much better than Fig. 3.14. For that we use the following special case of the <u>theorem of Ptolomy</u> :

In an isosceles trapezium, the product of the diagonals is equal to the sum of the products of the opposite sides.

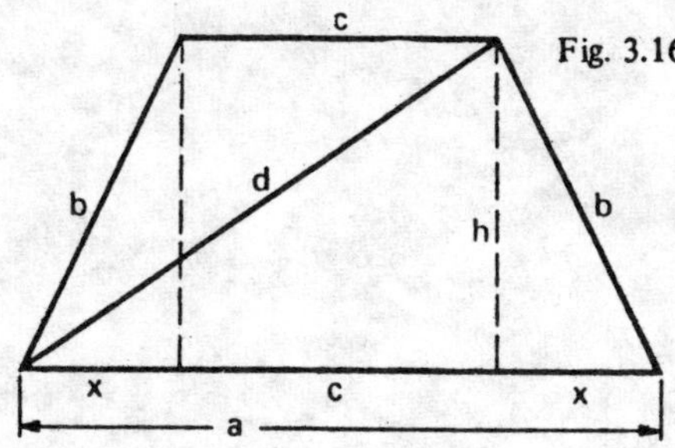

Fig. 3.16

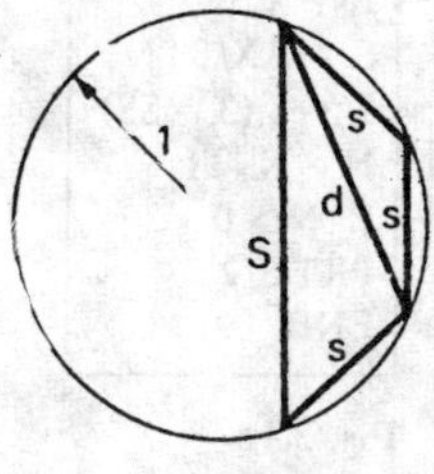

Fig. 3.17

<u>Proof</u>: In Fig. 3.16, from the theorem of Pythagoras we have,

$$h^2 = d^2 - (x + c)^2 = b^2 - x^2$$

$$d^2 = b^2 + (x + c)^2 - x^2 = b^2 + (x + c + x)(x + c - x)$$

$$d^2 = b^2 + ac \qquad \text{q.e.d.}$$

Fig. 3.17 gives

$$d^2 = s^2 + Ss.$$

On the other hand, from (1)

$$d^2 = s^2 (4 - s^2).$$

Equating these we have

(2) $S = 3s - s^3$

From this we obtain the program in Fig. 3.18. First the arc $2x$ is divided into 3^n equal small arcs. The chord of such a small arc approximates very closely to the curve itself, so we set the initial value of s to be $\frac{2x}{3^n}$. The substitution $s \leftarrow s(3 - ss)$ gives a chord which subtends an angle at the centre of the circle which is three times as big. After n steps we have the chord corresponding to the arc of length 2x. Halving this chord gives $\sin x$. Fig. 3.18, in contrast to Fig. 3.14, involves no square roots and needs a smaller value of n. For $n = 9$ the error is less than 10^{-10}. This is better than $n = 14$ in Fig. 3.14, since $3^9 = 19683$ and $2^{14} = 16384$. The program can be used with a pocket calculator without a square root facility.

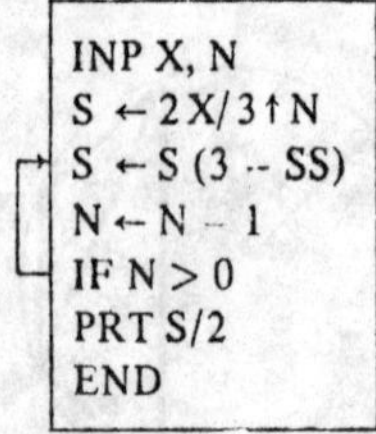

```
INP X, N
S ← 2X/3↑N
S ← S(3 - SS)
N ← N - 1
IF N > 0
PRT S/2
END
```

Fig. 3.18

Exercises

1. Try out the program in Fig. 3.18 by calculating $\sin\frac{\pi}{6} = 0.5$, $\sin\frac{\pi}{2} = 1$, $\sin\frac{\pi}{180} = 0.0174524064$ for $n = 1, 2, 3, \ldots$.. Which value of n is sufficient for 10-place accuracy?

2. Use the program in Fig. 3.10 to calculate
   a) $\arcsin x$ for $x = \frac{1}{2}, \sqrt{\frac{3}{2}}, 1$
   b) $\arccos x$ for $x = \frac{1}{2}, \sqrt{\frac{3}{2}}, 1$
   c) $\arctan x$ for $x = 1, \sqrt{3}, 10^{10}$

3. Show that, as for S, $\frac{T_{n+1} - T_n}{T_n - T_{n-1}} \longrightarrow \frac{1}{4}$ as $n \longrightarrow \infty$

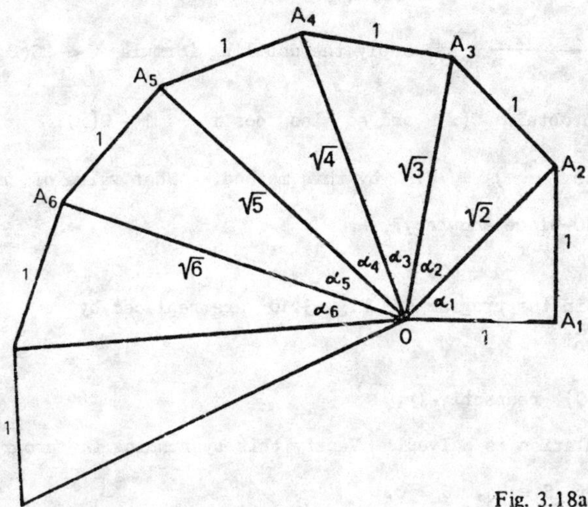

Fig. 3.18a

4. The polygonal path $A_1 A_2 A_3 A_4 \ldots$ in Fig. 3.18a is called a <u>square-root spiral</u>. Clearly $\overline{OA_n} = \sqrt{n}$, $\tan \alpha_n = \frac{1}{\sqrt{n}}$,

$\alpha_n = \arctan \frac{1}{\sqrt{n}}$.

   a) For which values of n does the spiral complete 1, 2, 3 .... 10 revolutions?

   b) Verify that $\overline{OA_n}$ increases by about π with each revolution.

5. All the formulas of 3.2 and 3.3 can be rephrased in trigonometrical language. In Fig. 3.2 let $\angle AMC = \alpha$.

   a) Find s_n, c_n, t_n, S_n, T_n.

   b) Show that the formulas (1) and (2) in 3.3 are equivalent to

   $\sin 2x = 2\sin x \cos x$, $\sin 3x = 3 \sin x - 4 \sin^3 x$.

6. We seek a program for cos x which is based on the double-angle formula $\cos 2x = 2\cos^2 x - 1$. For small $x (x < \frac{1}{2^{16}})$ we may take $\sin x \doteq x$ and

$$\cos x = 1 - 2\sin^2 \frac{x}{2} \doteq 1 - \frac{x^2}{2}.$$

To avoid error effects due to subtraction we write

$C(x) = 1 - \cos x \doteq \frac{x^2}{2}$.

a) Show that $C(2x) = 2\,C(x)(2 - C(x))$.

b) If we put $C \leftarrow \frac{x^2}{2.4^n}$ and apply the doubling formula $C \leftarrow 2C(2 - C)$ n times, we obtain $C(x)$ and so also $\cos x = 1 - C(x)$.

Calculate $\cos(\frac{\pi}{3}) = 0.5$ by this method. What value of n is needed for 10-place accuracy?

7. If lines 4 and 5 in the program of Fig. 3.10 are replaced by

IF $1 - C > 3.10^{-5}$

and PRT 3S/(2 + C) respectively,

the work of calculation is halved. Verify this by running the program in Fig. 3.18b for S = 2 , C = 0.

```
   INP S, C
┌→ C ← √((1 + C)/2)
│  S ← S/C
│  PRT 3S/(2 + C)
└─ IF 1 − C > 3 · 10⁻⁵
   END
```

Fig. 3.18b

(This may be proved by expanding the expression

$$\frac{3S_n}{2 + c_n} = \frac{3.2^n \sin \frac{\alpha}{2^n}}{2 + \cos \frac{\alpha}{2^n}}$$, where α = arc sin x , in an infinite series.)

3.4 The method of Cusanus

Nicolaus Cusanus, a philosopher of the Middle Ages, in 1450 discovered an elegant method of calculating π. Archimedes used a fixed circle and series of inscribed and circumscribed regular polygons of 3.2^n sides ($n = 1, 2, 3, \ldots$). Cusanus on the other hand used the sequence of regular 2^n - gons ($n = 2, 3, 4, \ldots$) with the <u>fixed perimeter</u> 2. Let h_n and r_n be the radii of the inscribed and circumscribed circles of each 2_n-gon in the sequence. These circles have circumference $2\pi h_n$ and $2\pi r_n$ respectively. The perimeter of the 2^n- gon lies between these (Fig. 3.19)

i.e. $$2\pi h_n < 2 < 2\pi r_n$$

$$\frac{1}{r_n} < \pi < \frac{1}{h_n}$$

Let $n = 2$, i.e. $2^n = 4$. For the square with perimeter 2 we have

$$r_2 = \frac{\sqrt{2}}{4}, \quad h_2 = \frac{1}{4}$$

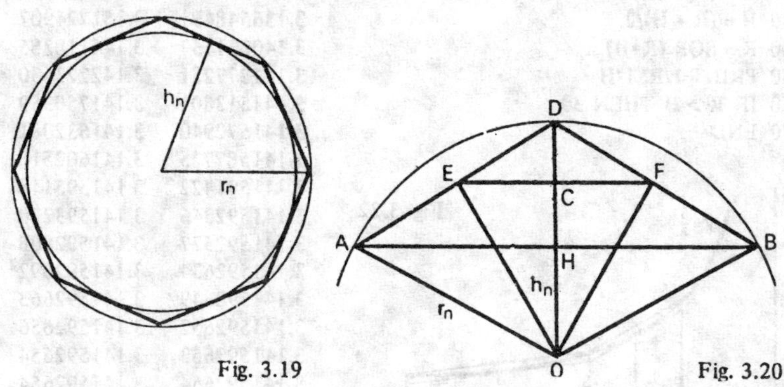

Fig. 3.19

Fig. 3.20

The Figure 3.20 gives recursion equations for r_n and h_n. AB is the side of a regular 2^n-gon with perimeter 2 and centre O. D is the mid-point of the arc AB, E and F the mid-points of the sides AD and BD in the triangle ABD. Thus $\overline{EF} = \frac{\overline{AB}}{2}$. Therefore EF is the side of a regular 2^{n+1}-gon with perimeter 2 and centre O. Thus

$$\overline{OD} = r_n, \quad \overline{OH} = h_n, \quad \overline{OE} = r_{n+1}, \quad \overline{OC} = h_{n+1}.$$

Since C is the mid point of DH, we have

$$h_{n+1} = \frac{r_n + h_n}{2}$$

In the right angled triangle ODE we have $\overline{OE}^2 = \overline{OC} \cdot \overline{OD}$ (by similar triangles)

i.e. $$r^2_{n+1} = r_n h_{n+1}.$$

Hence we have

$$h_2 = \frac{1}{4}, \quad r_2 = \frac{\sqrt{2}}{4}, \quad h_{n+1} = \frac{r_n + h_n}{2}, \quad r_{n+1} = \sqrt{r_n h_{n+1}}$$

It is very convenient that the formula for r_{n+1} uses the value h_{n+1} instead of h_n. On that account we can apply the substitutions $h \leftarrow \frac{r+h}{2}$, $r = \sqrt{rh}$ consecutively. The BASIC program in Fig. 3.21 needs no explanation.

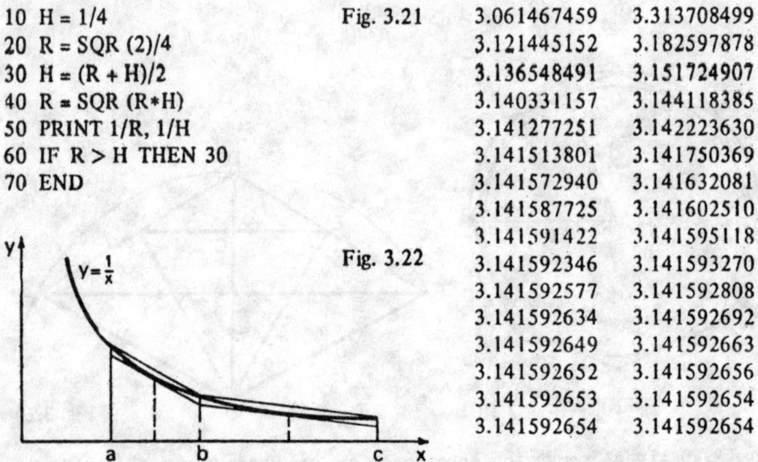

```
10 H = 1/4
20 R = SQR (2)/4
30 H = (R + H)/2
40 R = SQR (R*H)
50 PRINT 1/R, 1/H
60 IF R > H THEN 30
70 END
```

Fig. 3.21

3.061467459	3.313708499
3.121445152	3.182597878
3.136548491	3.151724907
3.140331157	3.144118385
3.141277251	3.142223630
3.141513801	3.141750369
3.141572940	3.141632081
3.141587725	3.141602510
3.141591422	3.141595118
3.141592346	3.141593270
3.141592577	3.141592808
3.141592634	3.141592692
3.141592649	3.141592663
3.141592652	3.141592656
3.141592653	3.141592654
3.141592654	3.141592654

Fig. 3.22

3.5 Quadrature of the hyperbola

a) Fig. 3.22 shows the hyperbola $xy = 1$ and two chord-trapezia (i.e. circumscribed) with areas S_1 and S_2. It is easy to calculate that

(1) $$S_1 = \frac{1}{2}\left(\frac{b}{a} - \frac{a}{b}\right) = \frac{1}{2}\left(x - \frac{1}{x}\right), \text{ where } x = \frac{b}{a}.$$

S_1 depends only in the quotient $x = \frac{b}{a}$. Therefore $S_1 = S_2$ provided that $\frac{b}{a} = \frac{c}{b}$, or $b^2 = ac$. i.e. $b = \sqrt{ac}$.

If we draw the tangents to the hyperbola at $\frac{a+b}{2}$ and $\frac{b+c}{2}$, we obtain two tangent-trapezia (i.e. inscribed) with areas T_1, T_2. A short calculation gives

$$(2) \qquad T_1 = 2\,\frac{b-a}{b+a} = 2\,\frac{\frac{b}{a}-1}{\frac{b}{a}+1} = 2\,\frac{x-1}{x+1}, \text{ where } x = \frac{b}{a}.$$

T_1 depends similarly only on $x = \frac{b}{a}$, so that in the event that $b = \sqrt{ac}$ we have $T_1 = T_2$ as well as $S_1 = S_2$.

(b) For $x > 0$ we define three functions s, c, t by

$$(3) \qquad s(x) = \frac{1}{2}\left(x - \frac{1}{x}\right), \quad c(x) = \frac{1}{2}\left(x + \frac{1}{x}\right), \quad t(x) = \frac{s(x)}{c(x)} = \frac{x^2-1}{x^2+1},$$

It is easily shown that

$$(4) \qquad s(\sqrt{x}) = \frac{s(x)}{2c(\sqrt{x})}, \quad c(\sqrt{x}) = \sqrt{\frac{1+c(x)}{2}}.$$

For $n = 0, 1, 2, 3, \ldots$ we define the sequence of numbers

$x_0 = x$, $x_1 = \sqrt{x}$, $x_2 = \sqrt{x_1} = \sqrt[4]{x}$, $x_3 = \sqrt{x_2} = \sqrt[8]{x}$, $x_4 = \sqrt{x_3} = \sqrt[16]{x}$, ... , and we write $s_n = s(x_n)$, $c_n = c(x_n)$, $t_n = t(x_n)$.

From $\sqrt{x_n} = x_{n+1}$ and using (4) we have

$$(5) \qquad s_{n+1} = \frac{s_n}{2c_{n+1}}, \quad c_{n+1} = \sqrt{\frac{1+c_n}{2}}, \quad t_{n+1} = \frac{s_{n+1}}{c_{n+1}}.$$

(c) Fig. 3.23 shows once more the hyperbola $xy = 1$.

For $x > 0$ we define the natural logarithm by

$\ln x$ = area under the hyperbola from 1 to x.

In order to calculate $\ln x$ we proceed by analogy with the circle problem. In Fig. 3.23 $\ln x$ is approximated by 2^n chord-trapezia each of the same area and with total area S_n, and by 2^{n-1} tangent trapezia likewise each of common area, whose total area is T_n. The equality of area of the trapezia is ensured by the relation

$$x_n^i = \sqrt{x_n^{i-1} x_n^{i+1}} = \sqrt{x_n^{i-2} x_n^{i+2}} .$$

Therefore we have the inequality $T_n < \ln x < S_n$. From (1) and (2) the area of the first chord-trapezium and of the first tangent trapezium are respectively

$$\frac{1}{2}\left(x_n - \frac{1}{x_n}\right) = s_n , \text{ and } 2\,\frac{x_n^2 - 1}{x_n^2 + 1} = 2t_n .$$

Also

(6) $$S_n = 2^n s_n , \quad T_n = 2^{n-1} \cdot 2t_n = 2^n t_n .$$

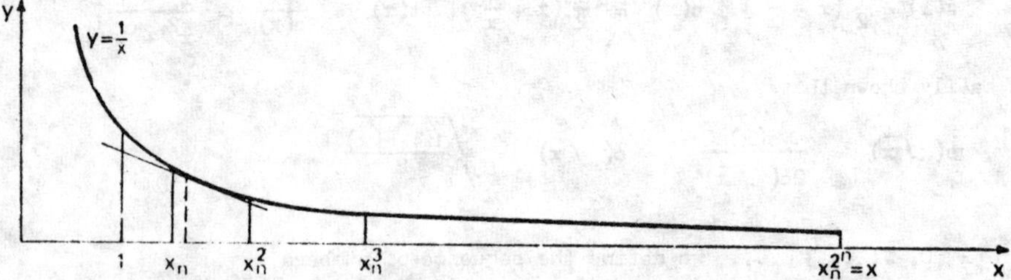

Fig. 3.23

From (5) and (6) we have

$$S_{n+1} = 2^{n+1} s_{n+1} = 2^{n+1} \frac{s_n}{2c_{n+1}} = \frac{S_n}{c_{n+1}} , \text{ and}$$

$$T_{n+1} = 2^{n+1} \frac{s_{n+1}}{c_{n+1}} = \frac{S_{n+1}}{c_{n+1}} ,$$

and hence the recursion equations

(7)
$$S_0 = \frac{1}{2}\left(x - \frac{1}{x}\right), \quad c_0 = \frac{1}{2}\left(x + \frac{1}{x}\right)$$
$$c_{n+1} = \sqrt{\frac{1+c_n}{2}} , \quad S_{n+1} = \frac{S_n}{c_{n+1}} , \quad T_{n+1} = \frac{S_{n+1}}{c_{n+1}}$$

Apart from the initial values, (7) and (5) in 3.2 correspond exactly, so we have shown the complete analogy with the circle problem.

For the circle, c_n converges to 1 from below. We shall show that for the hyperbola c_n converges to 1 from above. In fact, for $x \neq 1$, it is clear that

$$x_n = \sqrt{\sqrt{\ldots\sqrt{\sqrt{x}}}} \neq 1. \quad \text{Thus}$$

$$c_n = \frac{1}{2}\left(x_n + \frac{1}{x_n}\right) = 1 + \frac{1}{2}\left(\sqrt{x_n} - \frac{1}{\sqrt{x_n}}\right)^2 > 1, \quad \text{(strictly).}$$

For $n \to \infty$, $x_n \to 1$ and so also $c_n \to 1$. The sequences c_n, S_n, T_n, $S_n - T_n$ all have the convergence factor $\frac{1}{4}$. The proof of p. 74 applies without any alteration.

```
INP S, C
C ← √((1+C)/2)
S ← S/C
IF |1 − C| ≥ 10^-10
PRT S
END
```

Fig. 3.24

Input S	Input C	Output	Region of definition
x	$\sqrt{1-x^2}$	arc sin x	$-1 \leq x \leq 1$
$\sqrt{1-x^2}$	x	arc cos x	$-1 < x \leq 1$
$\frac{x}{\sqrt{1+x^2}}$	$\frac{1}{\sqrt{1+x^2}}$	arc tan x	$-\infty < x < \infty$
$\frac{1}{\sqrt{1+x^2}}$	$\frac{x}{\sqrt{1+x^2}}$	arc cot x	$-\infty < x < \infty$
$\frac{1}{2}(x - \frac{1}{x})$	$\frac{1}{2}(x + \frac{1}{x})$	ln x	$x > 0$
x	$\sqrt{1+x^2}$	ar sinh x	$-\infty < x < \infty$
$\sqrt{x^2-1}$	x	ar cosh x	$x \geq 1$
$\frac{x}{\sqrt{1-x^2}}$	$\frac{1}{\sqrt{1-x^2}}$	ar tanh x	$-1 < x < 1$
$\frac{1}{\sqrt{1-x^2}}$	$\frac{x}{\sqrt{1-x^2}}$	ar coth x	$-1 < x < 1$

Fig. 3.25

We can use the program of Fig. 3.10 to calculate ln x also, if we replace the exit condition by $|1 - C| < 10^{-10}$.

We summarise the preceding results as follows:

The "universal-program" in Fig. 3.24 calculates the new functions in table 3.25 with a relative error $\leqslant 10^{-10}$.

If one calculates arc cos x in the vicinity of -1, the result is inaccurate because of errors caused by subtraction.

We have proved four of the results in the table. The remaining ones are dealt with in the exercises.

A substantially improved version of this program is given in exercise 11.

(d) The function ln x is monotone increasing. Its inverse, $\ln^{-1} = \exp$ is called the exponential function . Fig. 3.26 shows the definition of ln and exp. If the area under the hyperbola from 1 to y has the value x, then y = exp x.†

To calculate y for a given value of x, we replace the area under the hyperbola from 1 to y by 2^n tangent-trapezia each with the area $\frac{x}{2^n}$.

In Fig, 3.27 they stretch from 1 to $y_n = y_1^{2^n}$. The tangent-trapezia lie below the curve, so that $y_n > \exp x$. However, the relative error is less than 10^{-10} for $n \geqslant 16$

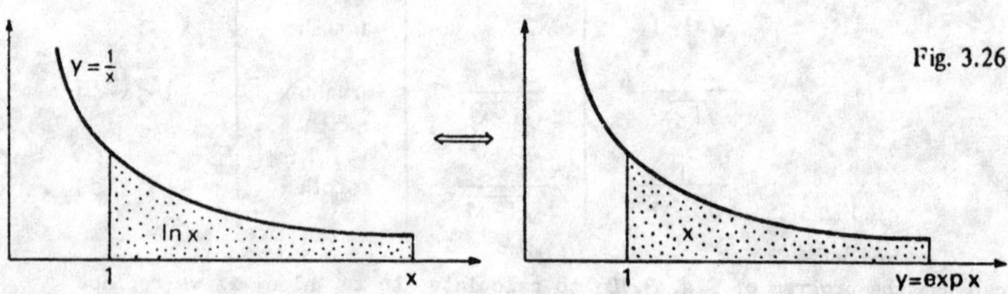

Fig. 3.26

† [Note that the axes in the L.H. diagram are labelled x, y. In the R.H. diagram the axes are u and v; x is the area under the curve and y = exp(x) a particular value of u]

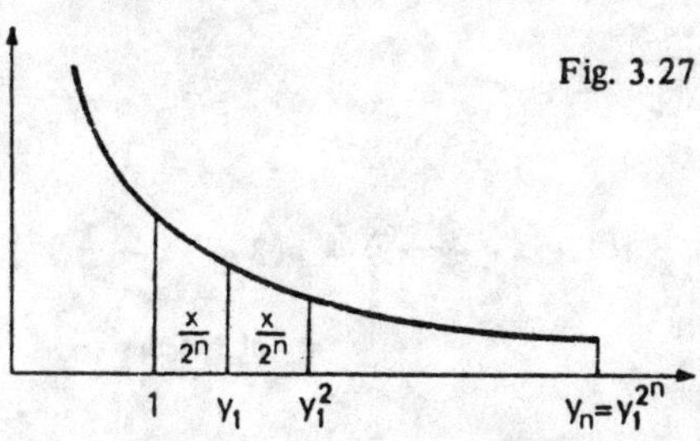

Fig. 3.27

From (2) the area of the first tangent-trapezium is given by

$$2\left(\frac{y_1 - 1}{y_1 + 1}\right) = \frac{x}{2^n} \Rightarrow y_1 = \frac{1 + \frac{x}{2^{n+1}}}{1 - \frac{x}{2^{n+1}}} \Rightarrow y_n = y_1^{2^n} = \left(\frac{1 + \frac{x}{2^{n+1}}}{1 - \frac{x}{2^{n+1}}}\right)^{2^n} \simeq \exp x.$$

i.e.

$$\text{(8)} \qquad \boxed{\exp x \simeq \left(\frac{1 + \frac{x}{2^{n+1}}}{1 - \frac{x}{2^{n+1}}}\right)^{2^n}}$$

The computer displays 10 decimal places, but internally works to 12 places.

The 13th and all subsequent places are rounded off. Therefore the initial value y_1 is subject to an error $\varepsilon \leqslant 5.10^{-12}$. What happens when one uses the square function 16 times?

Because $(1 + \varepsilon)^{2^{16}} \simeq 1 + 2^{16}\varepsilon$, and $2^{16}.\,\varepsilon \leqslant 0{\cdot}33.\ 10^{-6}$,

the rounding errors affect the preceding 5 or 6 decimal places. Thus the 6th place after the decimal point is unreliable. In fact

$$y_1^{2^{16}} = 2.718280760 \text{ instead of the correct answer } 2{\cdot}718281828.$$

In 3.6 we shall see how this rounding effect can be avoided. The exponential function will be considered further in Chapter 4.

(e) We construct a rapid exponental program, free from rounding effects, following the pattern of the sine program in Fig. 3.18. For this we need the analogue of the formula (2) given there. Again, we consider chord-trapezia inscribed in the hyperbola $xy = 1$. From (1), the chord trapezia from 1 to x and from 1 to x^3 have areas

$$s = \frac{1}{2}(x - \frac{1}{x})$$

and $$S = \frac{1}{2}(x^3 - \frac{1}{x^3}) = \frac{1}{2}(x - \frac{1}{x})(1 + x^2 + \frac{1}{x^2}) = s(3 + (x - \frac{1}{x})^2)$$

$$= s(3 + 4s^2)$$

so that $S = s(3 + 4s^2)$

The area under the hyperbola from 1 to exp x has the value x. We approximate this by 3^n equal-area chord-trapezia with total area x. In Fig. 3.28 they stretch from 1 to $y_n = y_1 \uparrow 3^n$. Since the chords lie above the hyperbola, $y_n < \exp x$.

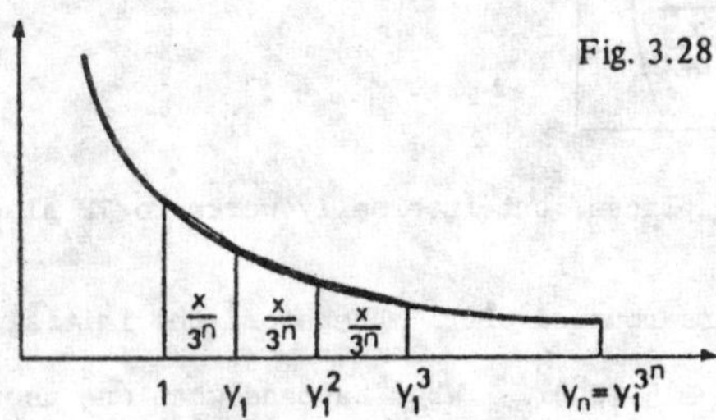

Fig. 3.28

For large n, however, we can write $y_n \doteq \exp x$. The chord-trapezium from 1 to y_1 has area $s = \frac{x}{3^n}$. By means of the substitution $s \leftarrow s(3 + 4s^2)$, we obtain the area of the (larger) chord-trapezium from 1 to y_1^3. After n substitutions we obtain the total area of the chord trapezia from 1 to y_n. Thus

$$\frac{1}{2}(y_n - \frac{1}{y_n}) = s \Rightarrow y_n = s + \sqrt{1 + s^2} \doteq \exp x.$$

From this we have the program in Fig. 3.29. This program was tested by calculating $e = \exp(1) = 2{\cdot}718281828$ for the input values n = 1 to n = 10.

```
INP X, N
S ← X/3↑N
S ← S (3 + 4SS)
N ← N - 1
IF N > 0
PRT S + √(1 + SS)
END
```

Fig. 3.29

n	e_n
1	2.670726281
2	2.712725190
3	2.717660819
4	2.718212782
5	2.718274156
6	2.718280976
7	2.718281733
8	2.718281818
9	2.718281827
10	2.718281828

Fig. 3.30

Exercises

1. Fig. 3.31 gives a "Universal algorithm" for calculating all the functions in Table 3.32. The algorithms can be verified by calculating $\ln 2 = 0.6931471806$, $\ln e = 1$, $2 \arcsin 1 = \pi$, $4 \arctan 1 = \pi$, $3 \arccos \frac{1}{2} = \pi$. The proofs of the correctness of the algorithms for ln and arc sin are sketched in exercises 2 and 3. The remaining statements then follow almost without calculation. A substantially improved version of the program is given in exercise 11.

```
INP A, B, C
A ← (A + B)/2
B ← √(AB)
IF A ≠ B
PRT C/A
END
```

Fig. 3.31

Input A	Input B	Input C	Output C/A	Region of definition
$(1+x)/2$	$\sqrt{x}$	$x-1$	$\ln x$	$x > 0$
$\sqrt{1-x^2}$	1	x	$\arcsin x$	$-1 \leqslant x \leqslant 1$
x	1	$\sqrt{1-x^2}$	$\arccos x$	$-1 < x \leqslant 1$
1	$\sqrt{1+x^2}$	x	$\arctan x$	$-\infty < x < \infty$
$\sqrt{1+x^2}$	1	x	ar sinh x	$-\infty < x < \infty$
x	1	$\sqrt{1+x^2}$	ar cosh x	$x \geqslant 1$
1	$\sqrt{1-x^2}$	x	ar tanh x	$-1 < x < 1$

Fig. 3.32

2. Let $a_1 = \frac{1+x}{2}$, $b_1 = \sqrt{x}$, $a_{n+1} = \frac{a_n + b_n}{2}$, $b_{n+1} = \sqrt{a_{n+1} b_n}$.

   Let S_n and T_n be defined as in 3.5.

   a) Show that $a_1 = \frac{x-1}{T_1}$, $b_1 = \frac{x-1}{S_1}$.

b) Show by mathematical induction that $a_n = \frac{x-1}{T_n}$, $b_n = \frac{x-1}{S_n}$

for all n. Hence

$$\lim_{n\to\infty} a_n = \lim_{n\to\infty} b_n = \frac{x-1}{\ln x}$$

3. In this exercise we use the notation of 3.2. See also Fig. 3.33.

Let $a_o = \sqrt{1-x^2}$, $b_o = 1$, $a_{n+1} = \frac{a_n + b_n}{2}$, $b_{n+1} = \sqrt{a_{n+1} b_n}$

a) Show that $a_1 = c_1^2 = \frac{2x}{T_1}$, $b_1 = c_1 = \frac{2x}{S_1}$.

b) Show by mathematical induction that $a_n = \frac{2x}{T_n}$, $b_n = \frac{2x}{S_n}$ for all n,

and hence that $\lim_{n\to\infty} a_n = \lim_{n\to\infty} b_n = \frac{x}{\text{arc sin } x}$.

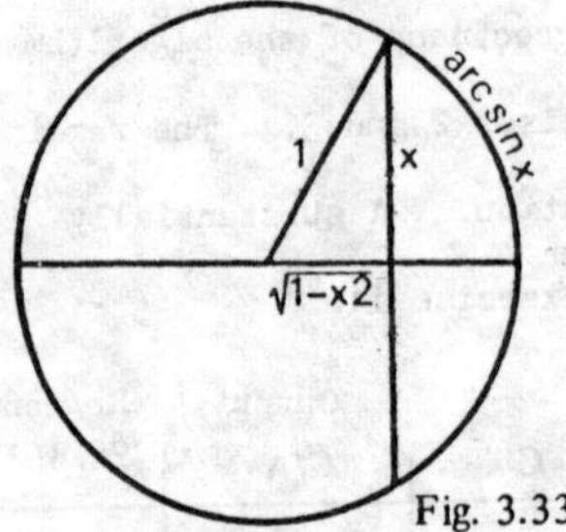

Fig. 3.33

4. What is the convergence factor of the algorithm in Fig. 3.31?

5. Fig. 3.31 gives inaccurate results for arc cos x in the neighbourhood of $x = -1$, because of subtraction errors. Check this.

6. The algorithm in Fig. 3.34 produces the output $\ln(1 + x)$ when x is input.

a) Check this by calculating ln 2.

b) By interchanging lines 5 and 6, the sequence of approximations may be printed out. What is the speed of convergence of this sequence?

```
INP X
A ← 1
X ← X/(√(1+X) + 1)
A ← 2A
IF X > 10^-10
PRT AX
END
```

Fig. 3.34

7. a) Let $s = \frac{1}{2}(x - \frac{1}{x})$, $c = \frac{1}{2}(x + \frac{1}{x})$, $S = s(x^2) = \frac{1}{2}(x^2 - \frac{1}{x^2})$.

Show that $c^2 - s^2 = 1$, $S = 2sc = 2s.\sqrt{1 + s^2}$

b) Write an exponential program analogous to that of Fig. 3.29, based on the doubling formula $S = 2s.\sqrt{1 + s^2}$, and so calculate the number e.

8. a) Write $G(x) = e^x - 1$ and show that $G(2x) = G(x)(2 + G(x))$.

b) An exponential program is to be constructed using the doubling formulae in (a). Note that $G(\frac{x}{2^n}) \simeq \frac{x}{2^n}$ for $n \geqslant 32$ with a relative error $\triangleq x.10^{-10}$. Use your program to calculate the number e.

9. We wish to re-examine the definitions of some of the functions in Figs. 3.25 and 3.32.

With $\exp x$ written as e^x, we define the so-called <u>hyperbolic functions</u>

$$\sinh x = \frac{e^x - e^{-x}}{2}, \quad \cosh x = \frac{e^x + e^{-x}}{2}, \quad \tanh x = \frac{\sinh x}{\cosh x}, \quad \coth x = \frac{\cosh x}{\sinh x}$$

(read as : hyperbolic sine of x, etc.). The inverses of these functions are the so-called <u>area-functions</u> [† in English usage inverse hyperbolic functions, $\sinh^{-1}$, etc.] (area-hyperbolic sine, etc.):

$$\text{arsinh } x = \ln(x + \ x^2 + 1),\ x \in \mathcal{R}, \quad \text{arcosh } x = \ln(x + \ x^2 - 1),\ x \geqslant 1$$

$$\text{artanh } x = \frac{1}{2}\ln\frac{1 + x}{1 - x},\ |x| < 1, \quad \text{arcoth } x = \frac{1}{2}\ln\frac{x + 1}{x - 1},\ |x| > 1.$$

Show that Fig. 3.24 outputs arsinh x when the inputs $S = x$, $C = \sqrt{1 + x^2}$ are used.

<u>Hint</u>: Let $u = x + \sqrt{x^2 + 1}$. Show that

$$s(u) = \frac{1}{2}(u - \frac{1}{u}) = x, \quad c(u) = \frac{1}{2}(u + \frac{1}{u}) = \sqrt{1 + x^2}$$

10. The formulae of 3.5 can be written in terms of the hyperbolic functions. If we write $x = e^t$, i.e. $t = \ln x$, then

$$s(x) = \frac{1}{2}(x - \frac{1}{x}) = \frac{e^t - e^{-t}}{2} = \sinh t = \sinh(\ln x),$$

†(translator's note)

$$c(x) = \frac{1}{2}\left(x + \frac{1}{x}\right) = \cosh t = \cosh(\ln x),$$

$$t(x) = \frac{s(x)}{c(x)} = \tanh t = \tanh(\ln x).$$

Calculate s_n , c_n , t_n , S_n , T_n in 3.5

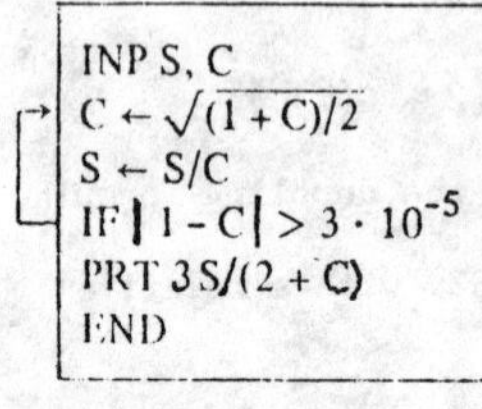

```
INP S, C
C ← √((1 + C)/2)
S ← S/C
IF |1 - C| > 3 · 10^-5
PRT 3S/(2 + C)
END
```

Fig. 3.35

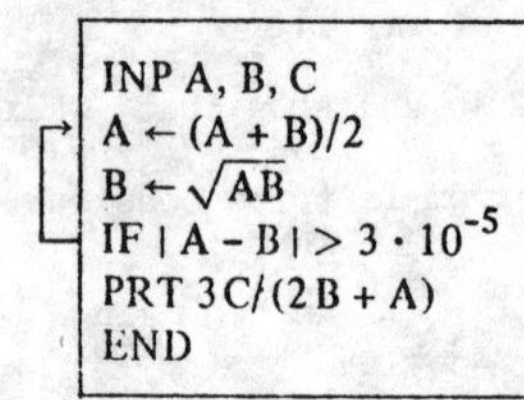

```
INP A, B, C
A ← (A + B)/2
B ← √(AB)
IF |A - B| > 3 · 10^-5
PRT 3C/(2B + A)
END
```

Fig. 3.36

11. Fig. 3.35 and 3.36 are considerable improvements on the universal programs of Fig. 3.24 and 3.31.

    Verify this by calculating $\ln 2 = 0{\cdot}6931471806$, $\ln e = 1$, $\arcsin 1 = 1{\cdot}570796327 = \frac{\pi}{2}$. It is worthwhile to interchange lines 4 and 5 so that the sequence of approximations is printed. How quickly does the sequence converge?

3.6 Accelerated convergence. The Romberg table.

The algorithms in 3.1 – 3.5 have convergence factor $\frac{1}{4}$. Christian Huygens made use of this with accelerated convergence in 1654. As an example we consider once more the algorithm in Fig. 3.37. It produces a sequence converging towards π , of terms S_o, S_1, S_2, ... At the same time (see Fig. 3.38)

$$S_o = \overline{AE}\ ,\ S_1 = \overline{AC} + \overline{CE}\ ,\ S_2 = \overline{AB} + \overline{BC} + \overline{CD} + \overline{DE}\ ,\ \ldots .$$

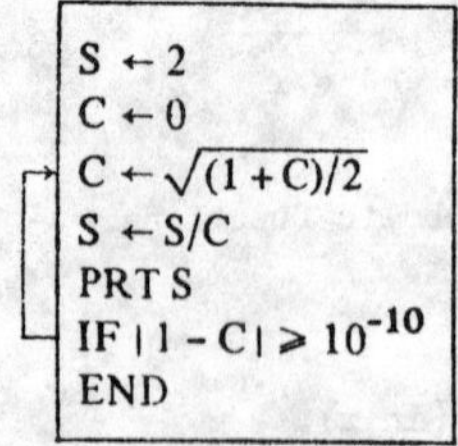

```
S ← 2
C ← 0
C ← √((1 + C)/2)
S ← S/C
PRT S
IF |1 - C| ≥ 10^-10
END
```

Fig. 3.37

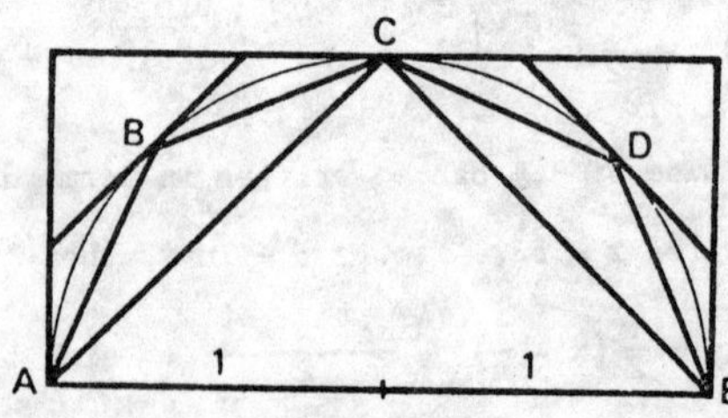

Fig. 3.38

The convergence factor is $\frac{1}{4}$. i.e. $S_n \simeq \pi + \frac{c}{4^n}$, $S_{n+1} \simeq \pi + \frac{c}{4^{n+1}}$

From this it follows that:

$$S'_n = \frac{4S_{n+1} - S_n}{3} \simeq \pi$$

It is therefore to be expected that S'_n will converge far faster than S_n.

Fig. 3.39 suggests that the sequence S'_n has convergence factor $\frac{1}{16} = 0{\cdot}0625$.

Only 8 iterations, instead of 16, are needed to obtain 10-place accuracy. The work of calculation is halved.

```
   S ← 2
   C ← 0
┌→ S1 ← S
│  C ← √((1+C)/2)
│  S ← S/C
│  PRT (4S − S1)/3
└─ IF |1 − C| ≥ 10^-5
   END
```

n	S'_n	$(S'_{n+1} - \pi)/(S'_n - \pi)$
0	3.104569500	0.0660
1	3.139147570	0.0634
2	3.141437717	0.0627
3	3.141582937	0.0626
4	3.141592046	0.0625
5	3.141592616	0.0624
6	3.141592651	
7	3.141592653	
8	3.141592654	

Fig. 3.39

Kepler and Simpson used the sequence S'_n to find the area under an arbitrary curve. (Simpson's Rule - see Chap. 4). Not until 1936 were the ideas of Huygens developed further, by Karl Kommerell. From

$$S'_n \simeq \pi + \frac{b}{16^n}, \quad S'_{n+1} \simeq \pi + \frac{b}{16^{n+1}}$$

it follows that

$$S''_n = \frac{16S'_{n+1} - S'_n}{15} \simeq \pi .$$

Consequently the sequence S''_n converges much more quickly than S'_n. Fig. 3.40 shows that it has convergence factor $\frac{1}{64}$. Only four iterations are needed for 10-figure accuracy. Analogously one may write

$$S'''_n = \frac{64S''_{n+1} - S''_n}{63} \qquad \text{with convergence factor } \frac{1}{256}\text{, and so on.}$$

Hence the so-called <u>Romberg table</u> can be displayed as follows:

S_0					
	S'_0				3.141452775
S_1		S''_0			3.141590393
	S'_1		S'''_0		3.141592618
S_2		S''_1		$S_0^{(4)}$	3.141592653
	S'_2		S'''_1		3.141592654
S_3		S''_2			Fig. 3.40
	S'_3				
S_4					

Each column in this table converges four times as fast as its predecessor. The table is easily constructed with a pocket calculator. Romberg introduced this summary table in 1955 for the determination of areas under arbitrary curves. For π one obtains the following table:

3.104569500			
	3.141452775		
3.139147570		3.141592578	
	3.141590393		3.141592654
3.141437717		3.141592653	
	3.141592618		
3.141582937			

The foregoing heuristic considerations need to be made more precise. This requires more knowledge — of trigonometry and of the series expansions of sin and tan.

$$\sin x = x - \frac{x^3}{3!} + \frac{x^5}{5!} - \frac{x^7}{7!} + \ldots,$$

$$\tan x = x + \frac{x^3}{3} + \frac{2}{15}x^5 + \frac{17}{315}x^7 + \ldots$$

For the length of the chords and tangents (S_n and T_n respectively in Fig. 3.29) we have

$$S_n = 2^n \cdot \sin\frac{\pi}{2^n}, \quad T_n = 2^n \cdot \tan\frac{\pi}{2^n},$$

$$S_n = \pi - \frac{\pi^3}{6\cdot 4^n} + \frac{\pi^5}{5!16^n} - \frac{\pi^7}{7!64^n} + \frac{\pi^9}{9!256^n} - \ldots \doteq \pi - \frac{\pi^3}{6\cdot 4^n}$$

$$T_n = \pi + \frac{\pi^3}{3\cdot 4^n} + \frac{2}{15}\frac{\pi^5}{16^n} + \frac{17}{315}\frac{\pi^7}{64^n} + \ldots \doteq \pi + \frac{\pi^3}{3\cdot 4^n}.$$

Thus $T_n - \pi \doteq 2(\pi - S_n)$. Consequently

$$U_n = \frac{2}{3} S_n + \frac{1}{3} T_n = \pi + \frac{\pi^5}{20 \cdot 16^n} + \dots$$

is a much better estimate for π and has convergence factor $\frac{1}{16}$.

For the justification of the Romberg table one needs only the fact that

$$(1) \qquad S_n = \pi + a_1 4^{-n} + a_2 4^{-2n} + a_3 4^{-3n} + \dots$$

Knowledge of the coefficients a_i is not necessary. From (1) it follows that

$$S_n' = \frac{4S_{n+1} - S_n}{3} = S_{n+1} + \frac{S_{n+1} - S_n}{3} = \pi + b_2 4^{-2n} + b_3 4^{-3n} + \dots$$

$$S_n'' = \frac{16S_{n+1}' - S_n'}{15} = S_{n+1}' + \frac{S_{n+1}' - S_n'}{15} = \pi + c_3 4^{-3n} + c_4 4^{-4n} + \dots$$

$$S_n''' = \frac{64S_{n+1}'' - S_n''}{63} = S_{n+1}'' + \frac{S_{n+1}'' - S_n''}{63} = \pi + d_4 4^{-4n} + d_5 4^{-5n} + \dots$$

$$\vdots$$

$$S_n^{(m)} = \frac{4^m S_{n+1}^{(m-1)} - S_n^{(m-1)}}{4^m - 1} = S_{n+1}^{(m-1)} + \frac{S_{n+1}^{(m-1)} - S_n^{(m-1)}}{4^m - 1} =$$

$$= \pi + e_{m+1} 4^{-(m+1)n} + e_{m+2} 4^{-(4+2)n} + \dots$$

The sequences S_n, S_n', S_n'', ... clearly have convergence factors $\frac{1}{4}$, $\frac{1}{16}$, $\frac{1}{64}$, ...

We now return to equation (8) in 3.5. There we had for $\exp x$ the approximation:

$$y_n(x) = \left(\frac{1 + \frac{x}{2^{n+1}}}{1 - \frac{x}{2^{n+1}}} \right)^{2^n}$$

From this $y_n(-x) = y_n(x)$, i.e. one can expand $y_n(x)$ as series with even powers of $\frac{x}{2^{n+1}}$:

$$y_n(x) = \exp(x) + c_2 x^2 . 4^{-n} + c_3 x^4 . 4^{-2n} + \ldots$$

By putting $x = 1$, and writing $y_n(1) = e_n$ we have, since $\exp(1) = e$,

$$e_n = e + c_2 4^{-n} + c_3 4^{-2n} + \ldots$$

Using $e_1 = (\frac{5}{3})^2$, $e_2 = (\frac{9}{7})^4$, $e_3 = (\frac{17}{15})^8$, $e_4 = (\frac{33}{31})^{16}$, we obtain the following Romberg table:

2.777777778			
	2.717555957		
2.732611412		2.718284237	
	2.718238719		2.718281827
2.721831893		2.718281865	
	2.718279168		
2.719167349			

The approximation 2.718281827 has absolute error $\leq 10^{-10}$ and relative error $< 0.5 \times 10^{-10}$.

Now we return to the calculation of ln by equation (7) of 3.5. Here we assume knowledge of the hyperbolic functions and their series expansion. If we write

$$x = \exp t, \quad \text{i.e.} \quad t = \ln x,$$

we obtain

$$c_o = \cosh t, \quad S_o = \sinh t, \quad c_n = \cosh \frac{t}{2^n}, \quad S_n = 2^n \sinh \frac{t}{2^n},$$

$$T_n = 2^n \tanh \frac{t}{2^n},$$

$$S_n = \ln x - \frac{\ln^3 x}{3!} 4^{-n} + \frac{\ln^5 x}{5!} 4^{-2n} - \frac{\ln^7 x}{7!} 4^{-3n} + \ldots$$

We put $x = 2$. The first column of the following table contains S_1, S_2, S_3, S_4.

0.7071067812			
	0.6931262723		
0.6966213995		0.6931471843	
	0.6931458773		0.6931471806
0.6940147578		0.6931471806	
	0.6931470992		
0.6933640138			

Hence $\ln 2 = 0{\cdot}6931471806$ in which all 10 digits are correct.

3.7 Lattice-points in a circle (π by counting)

Points with whole-number coordinates are called lattice-points. We consider all lattice points in the plane which obey the relation

(1) $\quad x^2 + y^2 < n$

These are all lattice points inside or on the circle centre the origin with radius $\sqrt{n}$. C. F. Gauss found a simple formula for the number $g(n)$ of such points. He divided them into 4 sets, A, B, C, D. A contains only the origin, and B those points on the axes apart from the origin. C contains the lattice points not on the axes and in or on the square of side $2\sqrt{\frac{n}{2}}$ inscribed in the circle (Fig. 3.41). D contains the remaining lattice points. The numbers of points in each set are

$$|A| = 1, \quad |B| = 4\left[\sqrt{n}\right], \quad |C| = 4\left[\sqrt{\frac{n}{2}}\right]^2, \quad |D| = 8\sum_{i=\left[\sqrt{\frac{n}{2}}\right]+1}^{\left[\sqrt{n}\right]} \left[\sqrt{n-i^2}\right]$$

(See Fig. 3.41). Hence

(2) $$g(n) = 1 + 4\left[\sqrt{n}\right] + 4\left[\sqrt{\frac{n}{2}}\right]^2 + 8\sum_{i=\left[\sqrt{\frac{n}{2}}\right]+1}^{\left[\sqrt{n}\right]} \left[\sqrt{n-i^2}\right].$$

The circle with radius $\sqrt{n}$ has area πn. Since $g(n) \simeq \pi n$,

(3) $$\pi \simeq \frac{g(n)}{n}$$

In spite of its unpleasant appearance, (2) is a suitable formula even for hand calculation. Only the whole number part of each square root expression is required, and this is easily estimated: For example

$$g(400) = 1 + 4.20 + 4.14^2 + 8\sum_{i=15}^{20} \left[\sqrt{400-i^2}\right]$$

$$= 1 + 80 + 784 + 8\ (13 + 12 + 10 + 8 + 6)$$

$$= 1257$$

$$\pi \approx \frac{g(400)}{400} = \frac{1257}{400} = 3.1425$$

Fig. 3.41

The program in Fig. 3.42 prints the table 3.43. For $n = 10^{10}$ one obtains the excellent approximation $\pi \doteq 3.1415925457$. Unfortunately, no precise statements can be made about the errors.

Let us write $g(n) = \pi n + f(n)$ where $f(n)$ is the error term, which is to be estimated. Around 1800 Gauss found the following coarse estimate

```
 10 FOR I = 1 TO 10
 20      G = 0
 30      N = 10↑I
 40      A = INT (SQR (N/2))
 50      B = INT (SQR (N))
 60      FOR J = A + 1 TO B
 70           G = G + INT (SQR (N - J * J))
 80      NEXT J
 90      G = 8 * G + 1 + 4 * B + 4 * A * A
100      PRINT N, G
110 NEXT I
120 END
```

Fig. 3.42

n	g (n)
10	37
100	317
1 000	3 149
10 000	31 417
100 000	314 197
1 000 000	3 141 549
10 000 000	31 416 025
100 000 000	314 159 053
1 000 000 000	3 141 592 409
10 000 000 000	31 415 925 457

Fig. 3.43

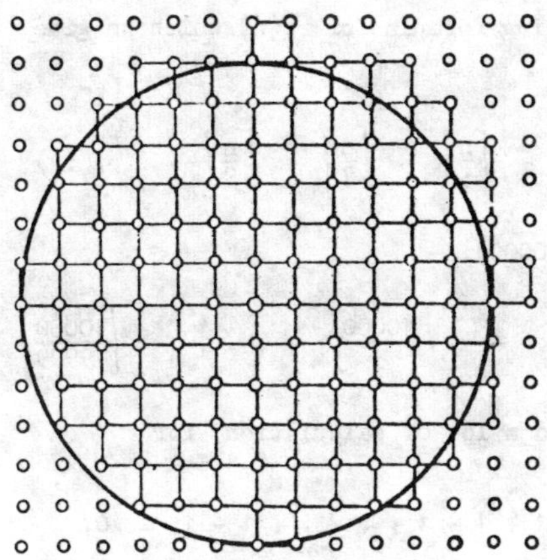

Fig. 3.44

With each lattice point in or on the circle he associated the unit square on its "North-East" side. The total area of these squares is $g(n)$. It does not correspond exactly to the area of the circle. Some squares extend outside the circle. On the other hand, there are regions inside the circle not covered by squares (Fig. 3.44).

However, all these squares lie within a circle with radius $\sqrt{n} + \sqrt{2}$ and a circle with radius $\sqrt{n} - \sqrt{2}$ is completely covered by the squares. Thus

$$\pi\,(\sqrt{n} - \sqrt{2})^2 < g(n) < \pi(\sqrt{n} + \sqrt{2})^2.$$

A simple rearrangement gives

(4) $$\left| g(n) - \pi n \right| < 2\pi(\sqrt{2n} + 1).$$

Instead of this we may write with a slight inaccuracy

(4) $$\left| g(n) - \pi n \right| < C_n \sqrt{n}$$

where C_n is bounded.

The estimates (4) and (4) are very coarse. In fact the error is much smaller. All that is known however is that

(5) $\left| g(n) - \pi n \right| < C_n n^{\alpha}$, with $\frac{1}{4} < \alpha < \frac{13}{40}$, where C_n is bounded.

Besides (2) there is another interesting formula for $g(n)$ which we give without proof;

$$(6) \qquad g(n) = 1 + 4\left([n] - \left[\frac{n}{3}\right] + \left[\frac{n}{5}\right] - \left[\frac{n}{7}\right] + \left[\frac{n}{9}\right] - \dots \right)$$

We use this formula to calculate $g(10000)$.

$$g(10000) = 1 + 4\left(\left[\frac{10000}{1}\right] - \left[\frac{10000}{3}\right] + \dots - \left[\frac{10000}{9999}\right] \right).$$

With a little forethought we can avoid a lot of calculation, for

$$\left[\frac{10000}{5001}\right] - \dots - \left[\frac{10000}{9999}\right] = 1 - 1 + 1 - 1 + \dots + 1 - 1 = 0,$$

and $\left[\frac{10000}{3337}\right] - \dots - \left[\frac{10000}{4999}\right] = 2 - 2 + 2 - 2 + \dots + 2 - 2 = 0,$

while $\left[\frac{10000}{2501}\right] - \dots - \left[\frac{10000}{3335}\right] = 3 - 3 + 3 - 3 + \dots + 3 - 2 = 1.$

Thus

$$g(10000) = 1 + 4\left(\left[\frac{10000}{1}\right] - \left[\frac{10000}{3}\right] + \left[\frac{10000}{5}\right] - \left[\frac{10000}{7}\right] + \dots - \left[\frac{10000}{2499}\right] + 1 \right).$$

Fig. 3.45 shows the corresponding BASIC program.

```
10 G = 0
20 FOR I = 1 TO 2497 STEP 4
30        G = G + INT (10000/I) - INT (10000/(I + 2))
40 NEXT I
50 PRINT 4 * G + 5
60 END

31417
```

Fig. 3.45

3.8 The Leibniz series for π

Leibniz showed that

(1) $$\frac{\pi}{4} = 1 - \frac{1}{3} + \frac{1}{5} - \frac{1}{7} + \frac{1}{9} - \dots$$

This famous series converges very slowly. Therefore it is particularly fruitful for teaching. For it can be used to test different methods of speeding up convergence. We can write (1) in two ways:

(2) $$\frac{\pi}{4} = (1 - \frac{1}{3}) + (\frac{1}{5} - \frac{1}{7}) + \dots = \frac{2}{1.3} + \frac{2}{5.7} + \frac{2}{9.11} + \dots$$

(3) $$\frac{\pi}{4} = 1 - (\frac{1}{3} - \frac{1}{5}) - (\frac{1}{7} - \frac{1}{9}) - \dots = 1 - \frac{2}{3.5} - \frac{2}{7.9} - \frac{2}{11.13} - \dots$$

In addition, we form the average of (2) and (3):

(4) $$\frac{\pi}{4} = \frac{1}{2} + \frac{4}{1.3.5} + \frac{4}{5.7.9} + \frac{4}{9.11.13} + \frac{4}{13.15.17} + \dots$$

Exercises

1) Sum 1000 terms of the series (1) and so obtain an estimate of π. The calculation should be done in four ways:

   (a) Add the terms successively from left to right.

   (b) Add from left to right, but keep the positive and negative terms separate and form the difference of the two sums.

   (c) Add the terms successively from right to left.

   (d) Add from right to left, but keep the positive and negative terms separate and subtract.

2) Find an estimate of π by summing 1000 terms of the series

   a) (2) b) (3) c) (4).

3) F. Vieta in 1593 found the first closed expression for π:

$$\frac{2}{\pi} = \frac{\sqrt{2}}{2} \cdot \frac{\sqrt{2 + \sqrt{2}}}{2} \cdot \frac{\sqrt{2 + \sqrt{2 + \sqrt{2}}}}{2} \cdot \dots$$

   Write a program which prints out 20 successive approximations to π.

4) In 1655 John Wallis found an infinite product for π:

$$\frac{\pi}{2} = \frac{2}{1} \cdot \frac{2}{3} \cdot \frac{4}{3} \cdot \frac{4}{5} \cdot \frac{6}{5} \cdot \frac{6}{7} \cdot \frac{8}{7} \cdot \frac{8}{9} \cdot \dots$$

   What approximation for π is given by the product of the first 1000 factors?

5) Euler showed that

a) $\frac{\pi^2}{6} = 1 + \frac{1}{2^2} + \frac{1}{3^2} + \frac{1}{4^2} + \ldots$

b) $\frac{\pi^4}{90} = 1 + \frac{1}{2^4} + \frac{1}{3^4} + \frac{1}{4^4} + \ldots$

What approximations for π are given by the first 1000 terms of these series?

3.9 Monte-Carlo method for determination of π.

The Monte-Carlo method involves the estimation of a number from the result of a probabalistic experiment. We give two examples.

(a) π from randomly-falling rain

Suppose 1000 raindrops fall randomly on the square in Fig. 3.46, and let the number of drops which fall inside the quarter circle be r. Then

$$\frac{r}{1000} \simeq \frac{\pi}{4} \quad \text{i.e.} \quad \pi \simeq \frac{r}{250}$$

Fig. 3.47 gives the program. The variable i counts all the "raindrops" and r counts those which fall in the quarter circle. The core of the program is line 30. We show how it works.

First a point $(x,y) = (\text{RND}, \text{RND})$ is chosen at random within the square. If the point lies inside the quarter circle, then $x^2 + y^2 < 1$, i.e. $\lfloor x^2 + y^2 \rfloor = 0$ and we have $r \leftarrow r + 1 - 0$ i.e. the drop is counted. If the point does not lie inside the quarter circle, then $1 < x^2 + y^2 < 2$ i.e. $\lfloor x^2 + y^2 \rfloor = 1$ and so $r \leftarrow r + 1 - 1$ i.e. the drop is not counted. The program gives poor estimates of π. It can be shown that

$$\pi - 0{\cdot}052 \leq \frac{r}{250} \leq \pi + 0{\cdot}052 \text{ with probability } 0{\cdot}683$$

and $$\pi - 0{\cdot}104 \leq \frac{r}{250} \leq \pi + 0{\cdot}104 \text{ with probability } 0{\cdot}955$$

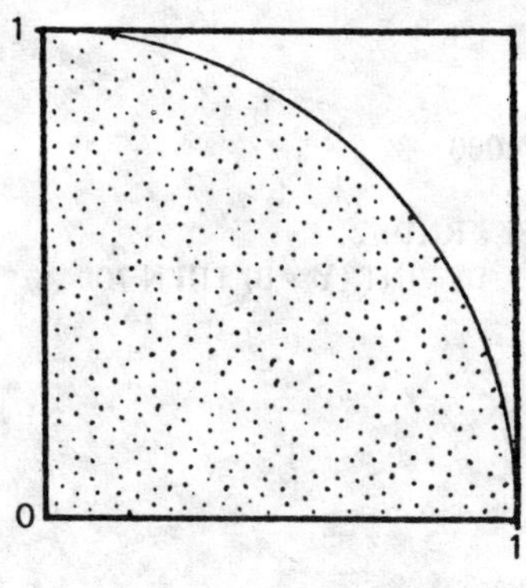

Fig. 3.46

```
10 R = 0
20 FOR I = 1  TO 1000
30        R = R + 1 – INT (RND↑2 + RND↑2)
40 NEXT I
50 PRINT R/250
60 END
```

3.104

Fig. 3.47

b) <u>Buffon's needle problem</u>

A table has parallel lines ruled across it at unit distance apart. We throw a needle of length 1, "at random" on the table.
How great is the probability that the needle falls across one of the lines? The answer is $\frac{2}{\pi}$. An outline of the proof is given in the appendix below. If the needle is thrown a large number of times and there are s crossings in w throws, then

$$\frac{s}{w} \doteq \frac{2}{\pi}, \quad \text{i.e.} \quad \pi \doteq \frac{2w}{s}$$

We write a program which produces an estimate of π by this method. All strips between the parallel lines are equivalent, so we need only consider that one in which the mid-point of the needle falls (Fig. 3.48). The abscissa of the mid-point is irrelevant. A 'throw' is obtained by first choosing a random ordinate y by $y \leftarrow$ RND, and then choosing the angle a randomly between 0 and π by $a \leftarrow \pi$. RND. If we write $b = 0{\cdot}5 \cos a$, then the needle cuts one of the parallel lines precisely when $[y_1] \neq [y_2]$, where $y_1 = y + b$, and $y_2 = y - b$. The program is given in Fig. 3.49.

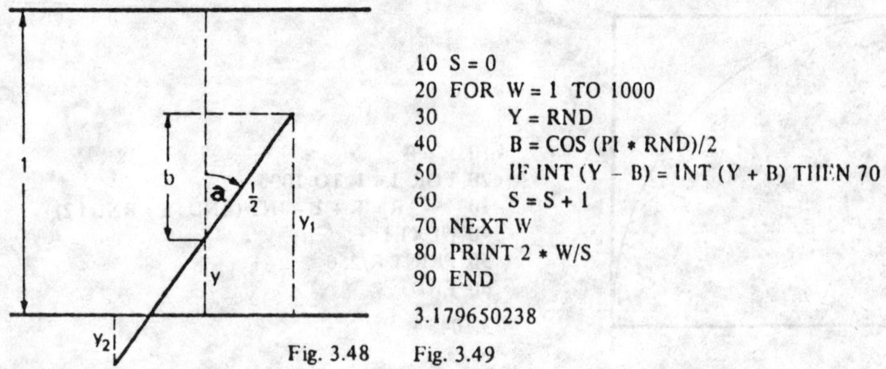


```
10 S = 0
20 FOR W = 1 TO 1000
30       Y = RND
40       B = COS (PI * RND)/2
50       IF INT (Y - B) = INT (Y + B) THEN 70
60       S = S + 1
70 NEXT W
80 PRINT 2 * W/S
90 END
3.179650238
```

Fig. 3.48 Fig. 3.49

Appendix

We throw at random onto the table a section of a curve of length L and arbitrary shape. (Fig. 3.50). We are interested in the average number f(L) of crossing points. "At Random" is taken to mean that any two short elements of the curve of equal length, have the same probability of containing a crossing point. It follows that f(L) is proportional to L and does not depend on the shape of the curve:

$$f(L) = L \cdot f(1).$$

If one chooses the circumference of a circle with diameter 1 and length $L = \pi$ this always has 2 crossing points (Fig. 3.51) i.e.

$$f(\pi) = \pi \cdot f(1) = 2 \Rightarrow f(1) = \frac{2}{\pi}.$$

i.e. with a needle of length 1 there are an average $\frac{2}{\pi}$ crossing points per throw. This is therefore the probability that the needle cuts one of the parallels. A detailed proof may be found in [5].

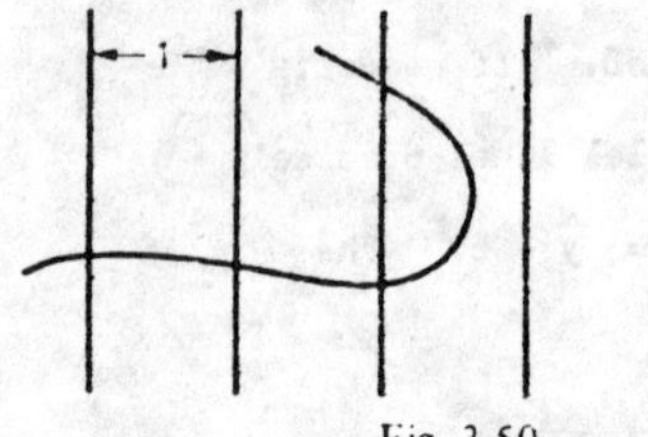

Fig. 3.50

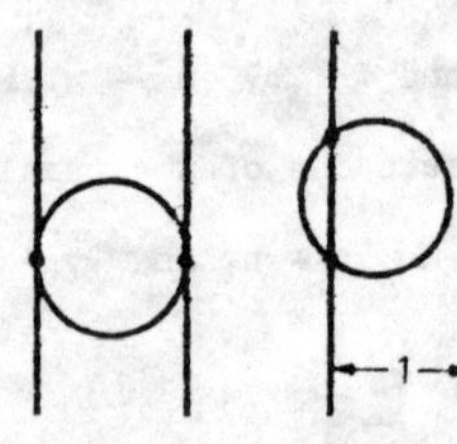

Fig. 3.51

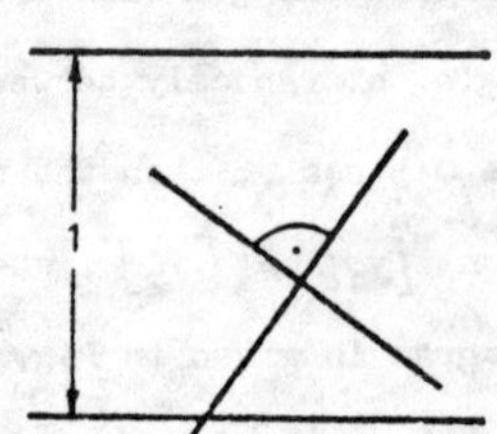

Fig. 3.52

Exercises

1) Better approximations for π are obtained by using 2 needles of length 1, which are fastened together at right angles, as in Fig. 3.52. Write a program which simulates 1000 throws of this cross.

2) A needle of length 1 is thrown at random on a lattice of squares of side 1. On s occasions in w throws there occurs at least one point of intersection. It can be shown that $\hat{\pi} = \frac{3w}{s}$ is a good approximation for π, and that one throw on a lattice of squares is equivalent to 12.08 throws on a 'striped' board. Write a program which simulates 1000 throws on a lattice of squares and so determines π using $\frac{3w}{s}$.

3) All points $(x_1, x_2, \ldots x_n)$ of n dimensional space with the property that $x_1^2 + \ldots + x_n^2 < 1$ form the interior of a ball centre 0 radius 1. Use 1000 randomly chosen points to estimate the content ('volume') of this ball for n = 3 and n = 4.

c) A game between Cain and Abel

The random number generator is an important source of intriguing data. The examination of such data leads to interesting and fruitful programs. As an example we consider the following game:

> Three segments A, B, C are each chosen at random from the interval (0, 1). If they can form the sides of a triangle, then Abel wins, otherwise Cain wins. The game is to be played 1000 times and from the data Abel's chance of winning is to be estimated.

The segments A, B, C form a triangle when one of the following conditions holds:

I $A + B > C$ and $A + C > B$ and $B + C > A$

II A + B + C > 2 MAX (A, B, C)

III $(A + B - C)(A + C - B)(B + C - A) > 0$

IV SGN (A + B - C) + SGN (A + C - B) + SGN (B + C - A) = 3

The segments do not form a triangle if:

V $A + B \leq C$ or $A + C \leq B$ or $B + C \leq A$.

The conditions I and V may be written respectively in BASIC as

I IF A + B > C AND A + C > B AND B + C > A THEN . . .

V IF A + B < = C OR A + C <= B OR B + C <= A THEN . . .

In IV the standard function SGN (signum) appears. It is defined by

$$\operatorname{sgn}(x) = \begin{cases} 1 & \text{for } x > 0 \\ 0 & \text{for } x = 0 \\ -1 & \text{for } x < 0. \end{cases}$$

We can use each of the five conditions as the basis of a program. However, it should be noted that in II, MAX* is not a standard function. Our program in Fig. 3.53 is particularly clear and instructive. S counts the games, and D counts the triangles (Abel's winning games). The whole of the work is done by line 60 which we will examine in detail. In BASIC each relation is evaluated and given the value 1 if it is true and 0 if it is false. If a triangle can be formed, each of the three brackets has the value 1 and D is increased by 1. If no triangle can be formed, then two of the brackets have the value 1 and one has value 0, i.e. D is unaltered.

```
10 D = 0
20 FOR S = 1 TO 1000
30        A = RND
40        B = RND
50        C = RND
60        D = D - 2 + (A + B > C) + (A + C > B) + (B + C > A)
70 NEXT S
80 PRINT D/1000
90 END

0.492
```

Fig. 3.53

We modify the game. The segments A, B, C form an obtuse-angled triangle, and acute-angled triangle, or no triangle at all. In the first case, Abel wins, in the second case Cain wins, and in the third case the result is a draw. The game is to be repeated 1000 times and a count made of the number of wins for Able and for Cain, and the number of draws. Here it is convenient to interchange the segments A, B, C so that C is the largest side. Then a triangle exists provided $C < A + B$, and if this relation is true, the winner is Abel or Cain according to whether $C^2 > A^2 + B^2$ or $C^2 \leq A^2 + B^2$ is true. In Fig. 3.54 the variables I, D, S are counters for all games, for triangles, and for obtuse-angled triangles (Abel's wins) respectively. (D-S) gives Cain's wins and I-D gives the number of drawn games. This program is written in an extended version of BASIC. It includes the exchange command "= = ", the IF-THEN-ELSE construction (line 60), and permits multiple statement lines.

```
 10 D = S = 0
 20 FOR I = 1  TO 1000
 30       A = RND; B = RND; C = RND
 40       IF A > B THEN A == B
 50       IF B > C THEN B == C
 60       IF C >= A + B THEN 80 ELSE D = D + 1
 70       IF C*C > A*A + B*B THEN S = S + 1
 80 NEXT I
 90 PRINT S, D - S, I - D
100 END

287            204            509
```

Fig. 3.54

Exercises

4) In each of the two squares in Fig. 3.55, a point is chosen at random. Let their distance apart be D. To calculate the average distance E(D) requires great labour. Therefore we wish to estimate E(D) by repeating the experiment 1000 times, determining the average of the distances. Write the corresponding program. (It can be shown that $E(D) \doteq 1.08814$).

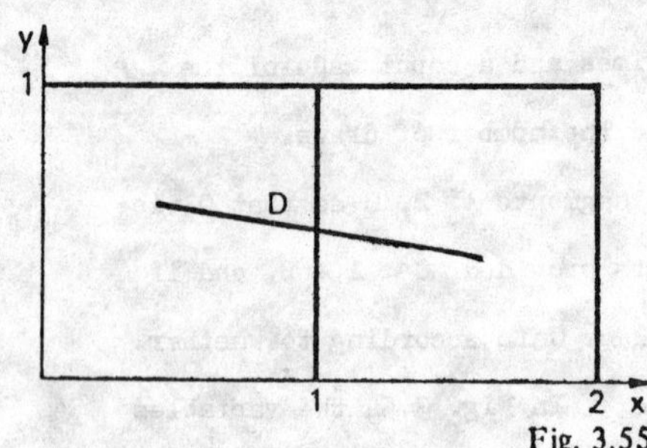

Fig. 3.55

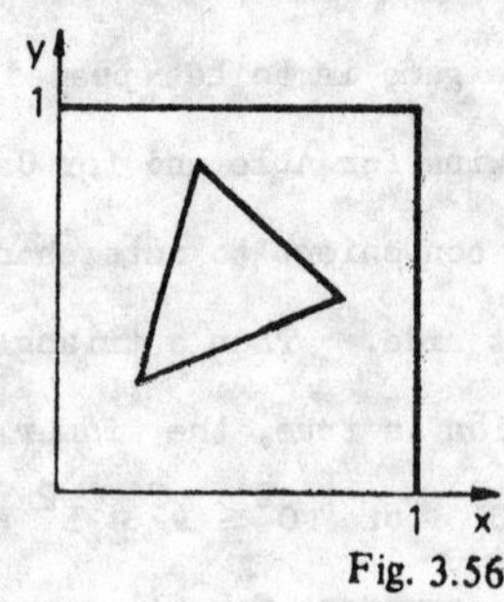

Fig. 3.56

5) Three points are chosen at random inside a unit square (FIg. 3.56). Let P be the probability that they form an obtuse-angled triangle. P is to be estimated by repeating the experiment 1000 times. (It can be shown, though with great labour, that $P = \frac{97}{150} + \frac{\pi}{40} \doteq 0.725$).

6) Two points are chosen at random in the interval (0,1). They divide the interval into 3 segments, A, B, C. Let P be the probability that the three segments form a triangle. Estimate P by repeating the experiment 1000 times.

4. NUMERICAL MATHEMATICS

4.1 Solution of an equation

A solution of the equation $f(x) = 0$ is also called a zero of f.

How should one find a zero of f?

a) Halving method [Bisection method]

We begin with a simple and reliable method which is particularly attractive for use with a computer. Suppose the function f is continuous in the interval $I = [a, b]$, and that $f(a)\,f(b) < 0$. i.e. that f has values of opposite sign at the endpoints of the interval. (Fig. 4.1) By the intermediate value theorem f has at least one zero in I.

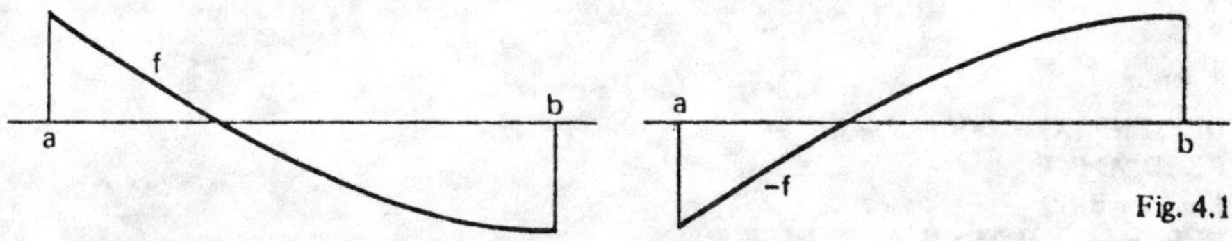

Fig. 4.1

I is the initial interval within which the zero is known to lie. We describe an algorithm which halves this interval at each step, until it is eventually smaller than some prescribed tolerance ε.

1. Put $x \leftarrow \frac{a + b}{2}$
2. If $f(x) = 0$, then STOP with x as answer.
3. If $f(x).f(a) > 0$ then put $a \leftarrow x$, otherwise put $b \leftarrow x$.
4. If $|a - b| > \varepsilon$ then go to step 1.
5. STOP with $x = \frac{a + b}{2}$ as answer.

After n steps the zero is determined with error $\leq \frac{(b - a)}{2^{n+1}}$.

We now assume that $f(a) < 0$ and $f(b) > 0$. (If not we multiply the equation $f(x) = 0$ throughout by -1). Fig. 4.2 shows a BASIC program. The IF-THEN-ELSE instruction is not possible (1976) on micro-computers. The unlikely event $f(x) = 0$ is not allowed for in this program. Line 10 requires some explanation. In BASIC about 10 standard functions, such as

SQR, INT, ABS, RND are provided. The user can define 26 further functions. A definition is introduced by the word DEF. Then comes FN, followed by one of the letters A to Z.

Line 10 defines a function which occurs several times in the program, and which can be called up by using its name FNF. If one wishes to use the same program with a different function then it is only necessary to retype line 10.

In Fig. 4.2 the three equations $x^2 - x - 1 = 0$, $x^3 - 2x - 5 = 0$, $x^2 - 4 \cos x = 0$ are solved, all of which have a zero in $[1, 3]$.

Warning! The computer determines $f(x)$ only approximately. If x is close to a zero, $f(x)$ may be so small that the calculated approximate value and the true value have opposite signs. The program does not provide for this.

```
10 DEF FNF (X) = X*X - X - 1
20 READ A, B, E
30 X = (A + B)/2
40 IF FNF (X) < 0 THEN A = X ELSE B = X
50 IF ABS (A - B) > E THEN 30
60 PRINT (A + B)/2
70 DATA 1, 3, 1E-10
80 END
   1.618033989
10 DEF FNF (X) = X*X*X - 2 * X - 5
   2.094551482
10 DEF FNF (X) = X*X - 4*COS (X)
   1.201538299
```

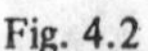

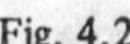

Fig. 4.2

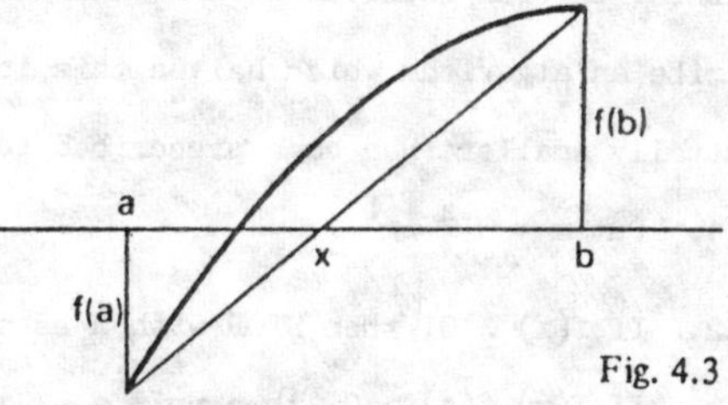

Fig. 4.3

b) The rule of false position (regula falsi).

Again suppose $f(a) < 0$ and $f(b) > 0$. We draw the chord joining the points $(a, f(a))$ and $(b, f(b))$. From Fig. 4.3 this intersects the x-axis at the point x where

$$\frac{x - a}{-f(a)} = \frac{b - x}{f(b)} \Rightarrow x = \frac{af(b) - b.f(a)}{f(b) - f(a)}.$$

If we replace $x \leftarrow \frac{a + b}{2}$ in the bisection algorithm by $x \leftarrow \frac{af(b) - bf(a)}{f(b) - f(a)}$, we obtain the regula falsi which in general is somewhat faster. (Fig. 4.4). The speed of convergence is not substantially greater than in the method of bisection. It can even be smaller, as Fig. 4.5 shows.

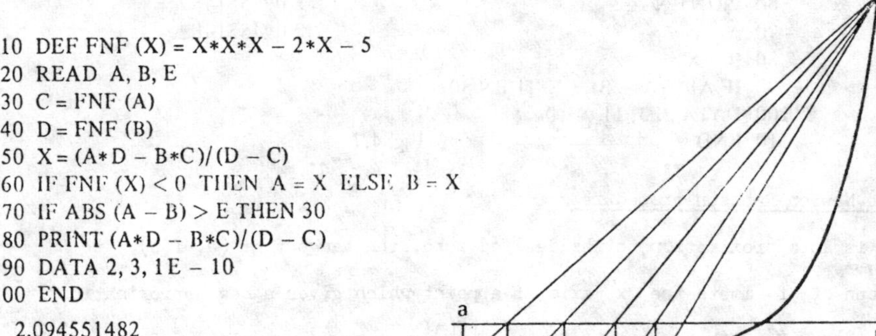

```
10 DEF FNF (X) = X*X*X - 2*X - 5
20 READ A, B, E
30 C = FNF (A)
40 D = FNF (B)
50 X = (A*D - B*C)/(D - C)
60 IF FNF (X) < 0 THEN A = X ELSE B = X
70 IF ABS (A - B) > E THEN 30
80 PRINT (A*D - B*C)/(D - C)
90 DATA 2, 3, 1E - 10
100 END
 2.094551482
```


Fig. 4.4

Fig. 4.5

c) The secant method

We calculate a sequence of approximations to the zero of f. From the two previous approximations a and b we calculate the next approximation

$$x = \frac{af(b) - bf(a)}{f(b) - f(a)}$$

The significance of x is clear from Fig. 4.6. The method is much quicker than regula falsi. However, it is less reliable, for x need not lie in $[a, b]$. The program in Fig. 4.7 calculates the zero of $x^3 - 2x - 5$ in only 5 iterations with an error of 10^{-10}. The regula falsi method required 19 iterations for this and the bisection method 33.

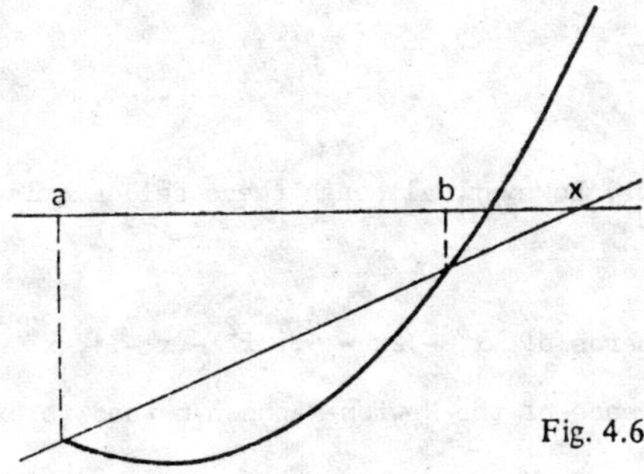

Fig. 4.6

```
 10 DEF FNF (X) = X*X*X - 2*X - 5
 20 READ A, B, E
 30 C = FNF (A)
 40 D = FNF (B)
 50 X = (A*D - B*C)/(D - C)
 60 PRINT X
 70 A = B
 80 B = X
 90 IF ABS (A - B) > E THEN 30
100 DATA 2, 3, 1E - 10
110 END
```

```
2.058823529
2.081263660
2.094824146
2.094549431
2.094551481
2.094551482
2.094551482
```

Fig. 4.7

d) The Newton-Raphson procedure

If a is an approximation to the desired zero, the tangent at (a, f(a)) to the graph of f meets the x axis at a point which gives a new approximation (Fig. 4.8). From the relation $f'(a) = \frac{f(a)}{a-x}$ we have, on solving, for x,

$$x = a - \frac{f(a)}{f'(a)}$$

Usually the Newton-Raphson method converges more quickly than the secant method.

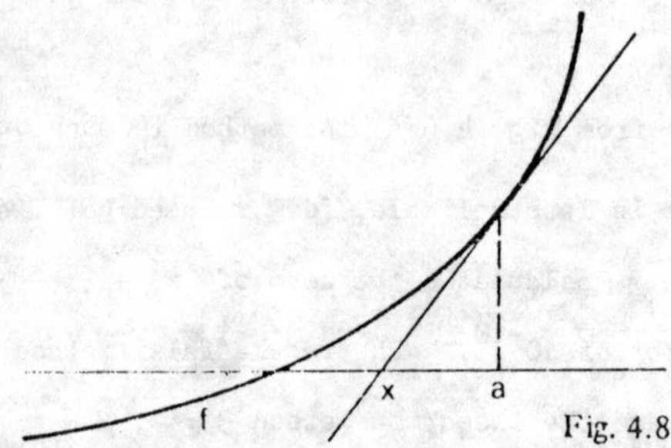

Fig. 4.8

It can break down in several ways, however, although these seldom arise in practice.

In Fig. 4.9 are determined the zeros of $x^3 - 2x - 5$, $x^2 - x - 1$, $x^2 - 4\cos x$ in $[1, 3]$. The rapid convergence of the Newton-Raphson method is examined in (e).

```
10 DEF FNF (X) = X*X*X - 2*X - 5
20 DEF FNG (X) = 3*X*X - 2
30 READ A, E
40 A = A - FNF (A)/FNG (A)
50 PRINT A
60 IF ABS (FNF (A)) > E THEN 40
70 DATA 2, 1E - 10
80 END
```

```
                  10 DEF FNF (X) = X*X - X - 1      10 DEF FNF (X) = X*X - 4*COS (X
                  20 DEF FNG (X) = 2*X - 1          20 DEF FNG (X) = 2*X + 4*SIN (X)

2.100000000       1.666666667                       1.258289035
2.094568121       1.619047619                       1.202379122
2.094551482       1.618034448                       1.201538498
2.094551482       1.618033989                       1.201538299
```

Fig. 4.9

Remark In the neighbourhood of a zero, the rounding errors in the calculation of $f(x)$ can be larger than the values of f itself, which must clearly be small. Too small a value of ε can produce an "infinite loop" where x oscillates about the zero. The error bound $\varepsilon = 10^{-10}$ which we have chosen, may be too small for some computers so that the program never terminates.

Rather than use $|A - B| < \varepsilon$ as a stop-condition it is better to use $\frac{|A - B|}{|A| + |B|} < \varepsilon$. The first makes the absolute, the second the relative, error less than ε. Suppose we test whether $|A - B| < 10^{-6}$. If the zero lies near 10^{-4}, it will be determined with relative error $\frac{10^{-6}}{10^{-4}} = 10^{-2}$ and in general this is too large an error. However, if the zero lies near 10^5, we require a relative error $\frac{10^{-6}}{10^5}$ or 10^{-11}. This is too exact, and the program will not stop.

Exercises.

1. Consider with the help of sketches how a) the secant rule, b) the Newton-Raphson method may fail.
2. Insert PRINT statements in Figs. 4.2 and 4.4 so that the whole sequence of approximations is printed.

3. Determine the real root of $x^3 - 3x - 4 = 0$ using each of the four methods a) to d). Calculate also the exact value

$$x = \sqrt[3]{2 + \sqrt{3}} + \sqrt[3]{2 - \sqrt{3}}.$$

4. Find the zeros of the following functions lying within the given intervals.

a) $y = x \ln x - 1$, $[1, 2]$ b) $y = x^3 - 10$, $[2,3]$

c) $y = x^3 - x - 1$, $[1, 2]$ d) $y = x^2 - \sin x$, $[0, 1]$

e) $y = x - \cos x$, $[0, 1]$ f) $y = x^5 + 5x + 1$, $[-1, 0]$.

5. We describe an algorithm for determining a zero of the continuous function f in $[a, b]$ when $f(a) < 0$ and $f(b) > 0$.

1. Start with $x = a$, put $h = 1$, and choose an error-bound ε.
2. Move to the right in steps of h until $f(x) > 0$.
3. Go back one step and print x.
4. Replace h by $\frac{h}{10}$. If $h \geqslant \varepsilon$, go back to step 2. Otherwise STOP.

Rewrite this algorithm in BASIC and determine the positive solutions of the following equations:

a) $x^3 - 2x - 5 = 0$ b) $x^5 - 10 = 0$ c) $x^7 - x - 1 = 0$

d) $x - \cos x = 0$ e) $x e^x - 1 = 0$.

6. Determine both solutions of the equation $xe^{-x} = \frac{1}{4}$ using Newton's method.

7. The equation $x^3 - 3x + 1 = 0$ has 3 real roots. Find them using Newton's method.

e) <u>Iteration</u>

We consider equations of the form

(1) $x = g(x)$, g continuous.

Equations of the form $f(x) = 0$ can be transformed into the form (1).

Suppose, for example, that $u(x) \neq 0$ for all x. Then

$$f(x) = 0 \iff u(x).f(x) = 0 \iff x = x + u(x).f(x)$$

We have transformed the equation $f(x) = 0$ into one of the form (1) with

(2) $g(x) = x + u(x).f(x)$.

The solutions of (1) are called <u>fixed points</u> of g. We look for a fixed point s of g. If we start with an approximation a for s, and use (1) systematically by substituting $x \leftarrow g(x)$, we obtain the program

(3)
```
  x <- a
> prt  x
  x <- g(x)
```

This program prints the sequence

(4) $$x_0 = a, \quad x_1 = g(x_0), \quad x_2 = g(x_1), \quad x_3 = g(x_2), \; . \; . \; .$$

If the sequence x_n converges to a limiting value s, then s is a fixed point of g. For, since g is continuous, from $x_{n+1} = g(x_n)$ it follows that

$$s = \lim_{n \to \infty} x_{n+1} = \lim_{n \to \infty} g(x_n) = g\left(\lim_{n \to \infty} x_n\right) = g(s)$$

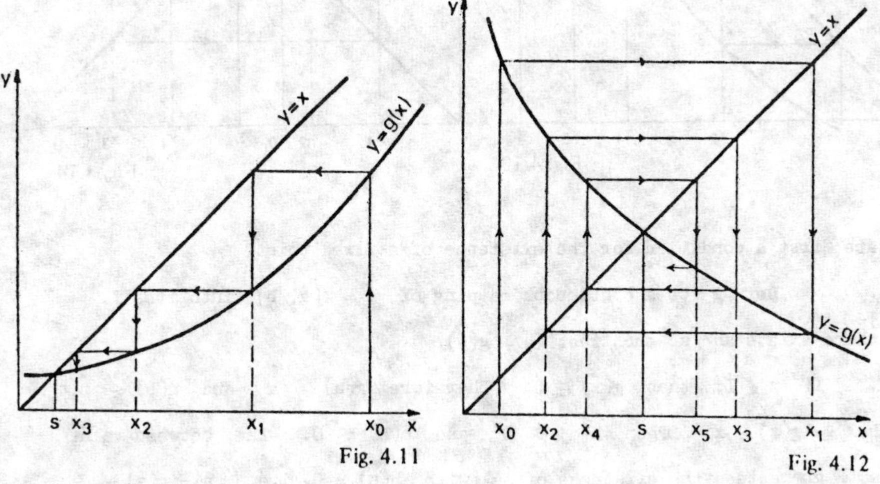

Fig. 4.11

Fig. 4.12

Fig. 4.11 shows the curves $y = x$ and $y = g(x)$. At their point of intersection $(s, g(s))$ we have $s = g(s)$. That is, s is a fixed point of g. The sequence x_n is easily constructed by means of the 'stair-case' diagram in Fig. 4.11. Here the sequence converges monotonically to s. In Fig. 4.12 the convergence of x_n is oscillatory . On the other hand,

Figs. 4.13 and 4.14 illustrate monotone and oscillating divergence respectively. If x_n converges, we have a method of calculating s in which the next approximation is found using the preceding ones. Such a method is called <u>iteration</u>, g is called an <u>iteration function</u> (iteration formula) and x_n is called an <u>iteration sequence</u>.

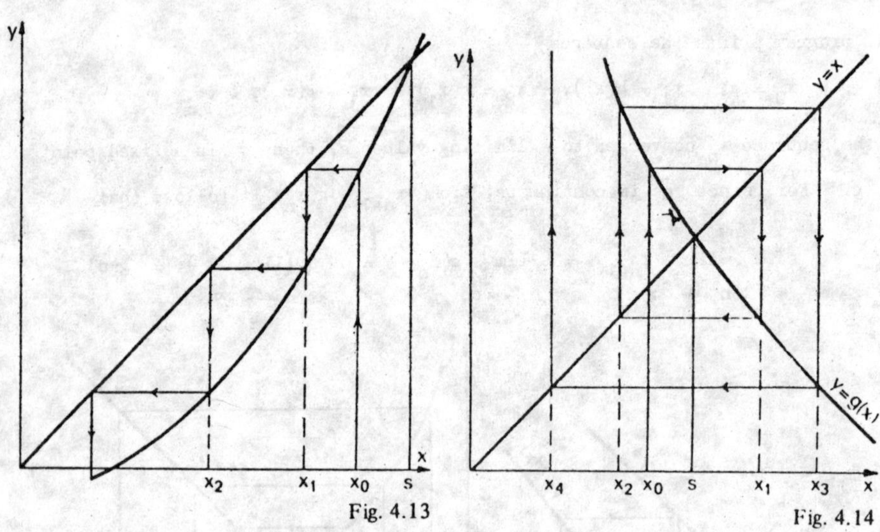

Fig. 4.13 Fig. 4.14

We state first a condition for the existence of a fixed point.

<u>Lemma</u>. Let g be a continuous mapping of $I = [a, b]$ into itself. Then there exists an s such that $s = g(s)$.

<u>Proof</u>. If $x \in I$, then $g(x) \in I$. Therefore $g(a) \geqslant a$ and $g(b) \leqslant b$. Let $h(x) = g(x) - x$. Then $h(a) \geqslant 0$ and $h(b) \leqslant 0$. The between-value theorem guarantees the existence of s with $h(s) = 0$, i.e. $s = g(s)$, which was to be proved.

We introduce a property of g which ensures the convergence of the sequence x_n.

<u>Definition</u> Let g be a mapping of the interval $I = [a, b]$ into itself. Then g is called a <u>contraction-mapping</u>, if there exists a constant L, $0 < L < 1$, such that

(5) $$|g(x) - g(y)| \leq L\,|x - y| \quad \text{for all } x, y \in I.$$

From (5) it follows that g is continuous. By the lemma, a contraction mapping $g: I \to I$ has at least one fixed point. We show that it is unique. Suppose there are two different fixed points s_1 and s_2. Then

$$s_1 = g(s_1), \quad s_2 = g(s_2), \quad s_1 - s_2 = g(s_1) - g(s_2).$$

From (5) it follows that,

$$|s_1 - s_2| = |g(s_1) - g(s_2)| \leq L|s_1 - s_2| < |s_1 - s_2|$$

and this is a contradiction.

We show now that the sequence $x_{n+1} = g(x_n)$, $n = 0, 1, 2, \ldots$ converges to the unique fixed point s, and in fact at least as quickly as the geometric sequence cL^n. This follows from

$$|x_n - s| = |g(x_{n-1}) - g(s)| \leq L|x_{n-1} - s| \leq L^n |x_0 - s|.$$

The condition (5) means, in geometrical terms, that every chord of the curve $y = g(x)$ has gradient $\leq L$, where $L < 1$.

The fixed point s of g is called an <u>attracting</u> fixed point if there is a neighbourhood $U = [s - a, s + a]$ of s, such that for all initial values in U the sequence $x_{n+1} = g(x_n)$ converges to s. If on the other hand there exists a sufficiently small neighbourhood of s, $U = [s - b, s + b]$ such that for every initial value x_0 ($x_0 \in U$, $x_0 \neq s$) the sequence $x_{n+1} = g(x_n)$ ultimately leaves U, we call s a <u>repelling</u> fixed point.

We have the following criteria for attracting and repelling fixed points:

Let g be a function continuous and differentiable $g: \mathbb{R} \to \mathbb{R}$ with $s = g(s)$. Then

$|g'(s)| < 1 \Rightarrow s$ is an attracting fixed point

$|g'(s)| > 1 \Rightarrow s$ is a repelling fixed point

$|g'(s)| = 1 \Rightarrow$ no general statement is possible.

<u>Proof</u>. Let a be so small that $|g'(x)| \leq L < 1$ in $U = [s - a, s + a]$

Then, for $x \in U$, the mean value theorem gives

$$|g(x) - g(s)| = |g(x) - s| = |g'(\xi)(x - s)| \leq L|x - s|$$

where ξ lies between x and s.

That is, g is a contraction mapping of U into itself, with fixed point s.

Now let $|g'(s)| > 1$. Choose b so that $|g'(x)| \geqslant \mu > 1$ in $U = [s - b, s + b]$. For $x \in U$ the mean value theorem gives

$$|g(x) - s| = |g(x) - g(s)| = |g'(\xi)(x - s)| \geqslant \mu |x - s|$$

That is, g is an expansion mapping and the sequence x_n diverges from s.

We now try to choose $u(x)$ in (2) in such a way that $g(x) = x + u(x).f(x)$ is as strongly contracting as possible in the neighbourhood of the fixed point. This is certainly the case when $g'(s) = 0$. Since $f(s) = 0$, we have $g'(s) = 1 + u(s).f'(s)$. The condition $g'(s) = 0$ is simply fulfilled if one chooses $u(x) = -\dfrac{1}{f'(x)}$.

This gives the Newtonian iteration function

$$(6) \qquad g(x) = x - \frac{f(x)}{f'(x)}$$

for the solution of the equation $f(x) = 0$. This choice of $u(x)$ leads to the iterative sequence

$$(7) \qquad x_{n+1} = x_n - \frac{f(x_n)}{f'(x_n)}$$

Remarks and Examples

1. Linear and superlinear convergence

Let s be an attracting fixed point of g and $x_{n+1} = g(x_n)$ with x_0 from the 'region of attraction' of s. We shall call $e_n = x_n - s$ the nth error. By the mean value theorem

$$e_{n+1} = x_{n+1} - s = g(x_n) - g(s) = g'(\xi)(x_n - s) = g'(\xi)\, e_n,$$

where ξ lies between x_n and s.

Letting $n \to \infty$, we have

$$\frac{e_{n+1}}{e_n} \to g'(s), \quad |g'(s)| = q < 1.$$

We say the sequence x_n converges linearly to s with convergence factor q, if

$$\lim_{n\to\infty} \frac{e_{n+1}}{e_n} = q, \quad 0 < |q| < 1.$$

For large n the errors e_n are nearly equal to the members of a geometric sequence with quotient q. If on the other hand, $g'(s) = 0$, then by Taylor's theorem

$$e_n = x_{n+1} - s = g(x_n) - g(s) = \frac{1}{2} g''(\xi)(x_n - s)^2 = \frac{1}{2} g''(\xi) e_n^2$$

$$\frac{e_{n+1}}{e_n^2} \rightarrow \frac{1}{2} g''(s) \quad \text{for} \quad n \rightarrow \infty .$$

Here we have <u>quadratic convergence</u> which is much more rapid than linear convergence. For from $|e_{n+1}| < |qe_n^2|$ it follows that $|qe_{n+1}| \leq |qe_n|^2$.

Thus if $|qe_n| < 10^{-m}$, then $|qe_{n+1}| < 10^{-2m}$.

Newton's method uses the iteration formula (6) to find a simple zero s of f. Then we have $f(s) = 0$ and $f'(s) \neq 0$ (simple zero). It follows that

$$g'(x) = \frac{f(x).f''(x)}{(f'(x))^2} , \text{ so that } g'(s) = 0.$$

That is, at a simple zero Newton's method gives quadratic convergence to s if one starts sufficiently close to s.

It can be shown that the secant method also has superlinear convergence (convergence faster than a geometric sequence). In fact

$$\frac{e_{n+1}}{e_n^{\phi}} \rightarrow q \quad \text{for} \quad n \rightarrow \infty \quad \text{where } \phi = \frac{\sqrt{5} + 1}{2} = 1.618 \ldots$$

2. A fixed point s of g is also a fixed point of the inverse function $h = g^{-1}$. If g is an expansion mapping, then h is a contraction mapping, since $h'(s) = \dfrac{1}{g'(s)}$. As an example, we find the smallest positive root of

(8) $\quad x = \tan x.$

Since $g'(x) = \tan' x = \dfrac{1}{\cos^2 x} > 1$, the sequence $x_{n+1} = \tan x_n$ diverges.

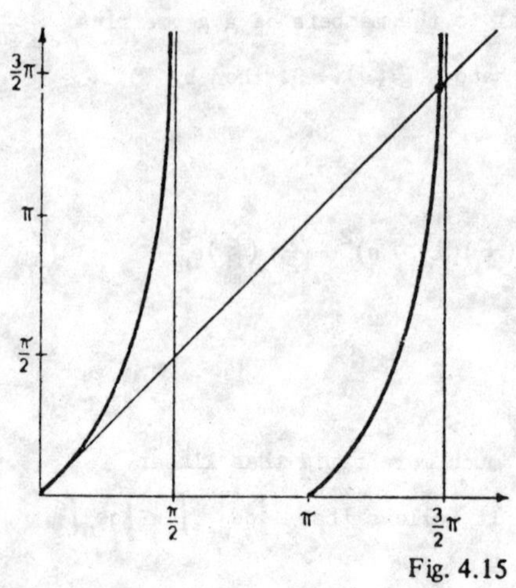

Fig. 4.15

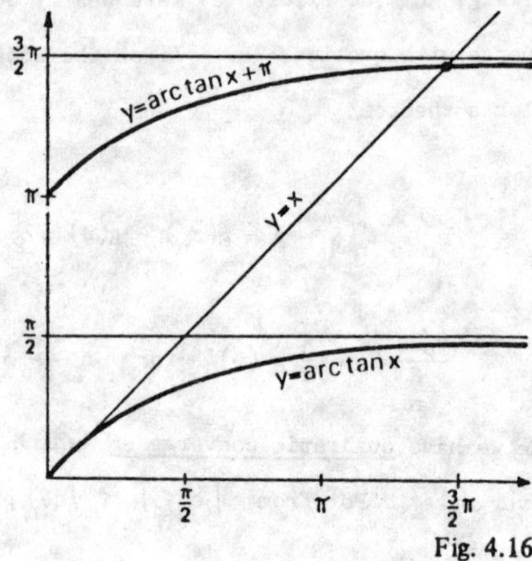

Fig. 4.16

If we reflect Fig. 4.15 in $y = x$ we obtain Fig. 4.16. Instead of (8) we solve

(9) $$x = \arctan x + \pi$$

Since $h'(x) = (\arctan x + \pi)' = \frac{1}{1+x^2} < 1$ the sequence

$$x_{n+1} = \arctan x_n + \pi \ , \quad x_0 = 4.5$$

converges to the desired fixed point s. Since $s \doteq 4.5$ and $h'(s) < \frac{1}{20}$, we have, for values close to s, $e_{n+1} < \frac{e_n}{20}$. One step of the iteration reduces the error about 20 fold. (Fig. 4.17)

```
10 READ B, E
20 A = B
30 B = ATN (A) + PI
40 PRINT B
50 IF ABS (A - B) > E THEN 20
60 DATA 4.5, 1E - 10
70 END
```

```
4.493720035
4.493424113
4.493410150
4.493409491
4.493409459
4.493409458
4.493409458
```

Fig. 4.17

3. Rate of convergence of Newton's method at a multiple zero

Let $f(x) = (x - a)^2$. Here $x = a$ is a double zero. The Newton iteration function is

$$g(x) = x - \frac{f(x)}{f'(x)} = x - \frac{x-a}{2} .$$

From this follows the iteration sequence

$$x_{n+1} = x_n - \frac{x_n - a}{2}$$

$$\text{i.e. } x_{n+1} - a = \frac{x_n - a}{2}$$

Thus $$x_n - a = \frac{x_0 - a}{2^n}$$

The convergence is linear with convergence factor $\frac{1}{2}$ i.e. no better than by the bisection method (Fig. 4.18).

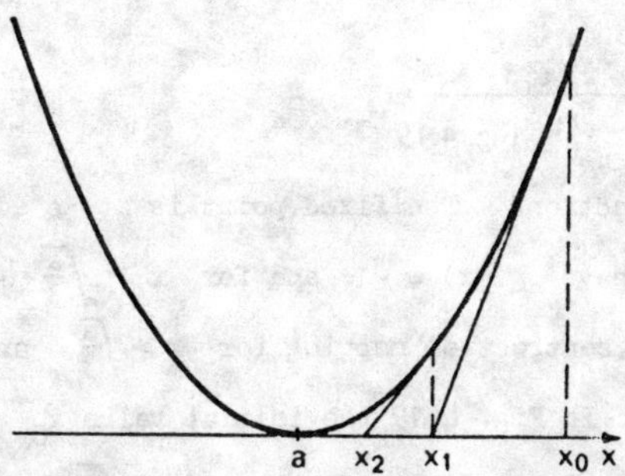

Fig. 4.18

Suppose now $f(x) = (x - a)^m$, so that a is an m-fold zero. After short calculation Newton's method gives

$$x_n - a = \left(\frac{m - 1}{m} \right)^n (x_0 - a)$$

The convergence is linear with the factor $q = \frac{m - 1}{m}$. For $m > 2$ this is worse than with the bisection method.

4. For $a > 0$, the function $f(x) = x^2 - a$ has a positive zero $\sqrt{a}$. The Newton iteration function (6) gives

$$(10) \qquad g(x) = \frac{1}{2} \left(x + \frac{a}{x} \right).$$

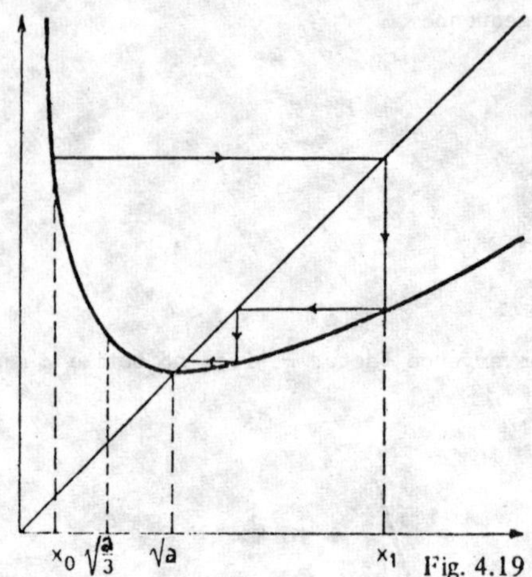

Fig. 4.19

Fig. 4.19 shows the graph of this function. The fixed point is at $s = \sqrt{a}$. There $g'(s) = 0$. At $x = \sqrt{\frac{a}{3}}$ we have $g'(x) = -1$ and for $x > \sqrt{\frac{a}{3}}$ we have $|g'(x)| < 1$. Thus g is a contraction mapping for $x > \sqrt{\frac{a}{3}}$ and an expansion mapping for $0 < x < \sqrt{\frac{a}{3}}$. In Fig. 4.19 the initial value x_0 is chosen in the expansion region. Therefore x_1 is further from $\sqrt{a}$ than is x_0. But x_1 lies in the contraction region. Hence we see that

$$(11) \qquad x_{n+1} = \frac{1}{2}\left(x_n + \frac{a}{x_n}\right)$$

converges to $\sqrt{a}$ for every positive initial value x_0. After x_1 the sequence decreases monotonically towards $\sqrt{a}$.

The iteration sequence (11) was used 4000 years ago to calculate square roots.

5. Exit conditions for iteration

In the floating point arithmetic of computers, the equation $A = B$ is seldom exactly satisfied. The sequence $x_{n+1} = f(x_n)$ converges for $n \to \infty$. It is incorrect to continue computing until $x_{n+1} = x_n$, since through the effects of rounding the sequence may be periodic with a very long period. It is correct to wait until $|x_{n+1} - x_n| < \delta$ for a suitable δ. However, since we do not know beforehand the order of magnitude of x_n, it is better still to compute until $|x_{n+1} - x_n| \leq \varepsilon x_n$ is satisfied. It is easier to make a more sensible choice of ε than of δ.

Exercises

8. In (11) let $e_n = x_n - \sqrt{a}$ be the nth absolute error.

(a) Show by algebraic manipulation that $e_{n+1} = \dfrac{e_n^2}{2x_n}$.

(b) Consider now, instead of the absolute error e_n, the relative error $\epsilon_n = \dfrac{e_n}{\sqrt{a}}$. Given that $x_n > \sqrt{a}$ for $n \geqslant 1$, show that

$$\epsilon_{n+1} < \frac{\epsilon_n^2}{2}.$$

9. In the squareroot algorithm (11) the value of x_n can be explicitly calculated.

Show first that

$$\frac{x_{n+1} - \sqrt{a}}{x_{n+1} + \sqrt{a}} = \left(\frac{x_n - \sqrt{a}}{x_n + \sqrt{a}}\right)^2$$

and hence that

$$\frac{x_n - \sqrt{a}}{x_n + \sqrt{a}} = \left(\frac{x_0 - a}{x_0 + a}\right)^{2^n}, \quad x_n = \sqrt{a}\,\frac{(x_0 + \sqrt{a})^{2^n} + (x_0 - \sqrt{a})^{2^n}}{(x_0 + \sqrt{a})^{2^n} - (x_0 - \sqrt{a})^{2^n}}$$

For $x_0 > 0$ we have $|x_0 + \sqrt{a}| > |x_0 - \sqrt{a}|$, so that $x_n \to \sqrt{a}$.

For $x_0 < 0$ we have $|x_0 + \sqrt{a}| < |x_0 - \sqrt{a}|$, so that $x_n \to -\sqrt{a}$.

10. For $a > 0$, the function $f(x) = x^m - a$ has a positive zero $\sqrt[m]{a}$.

Show that

$$(12) \quad x_{n+1} = \frac{1}{m}\left((m-1)x_n + \frac{a}{x_n^{m-1}}\right),$$

tends to $\sqrt[m]{a}$ for any $x_0 > 0$.

Calculate $\sqrt[3]{2}$, $\sqrt[10]{10}$, $\sqrt[5]{10}$ using a BASIC program.

11. Show that $g(x) = \sqrt{1 + x}$ is a contraction mapping for $0 < x < \infty$.

Determine the fixed point by iteration.

12. Show that $g(x) = 1 + \dfrac{1}{x}$ is a contraction mapping for $1 < x < \infty$.

Determine the fixed point by iteration.

13. Investigate the convergence of the sequence x_n defined by

$2x_{n+1} = \sqrt{1 + x_n}$ for different starting values.

14. Let $x_1 = 3$ and $x_{n+1} = \frac{4x_n + 10}{4 + x_n}$. Show that $x_n \rightarrow \sqrt{10}$ and

$|x_n - 10| < (\frac{1}{7})^n$.

15. Let $x_1 = \sqrt{a}$, $x_{n+1} = \sqrt{a + x_n}$, $a > 0$.

To what limit s does x_n converge. How large is s for $a = m(m + 1)$?

Give an explicit formula for x_n.

16. Find the limits of the sequences

a) $x_0 = 1$. $x_{n+1} = \frac{1}{1 + x_n}$ b) $x_0 = 1$, $x_{n+1} = 1 + \frac{1}{1 + x_n}$

17. The arithmetico-geometric mean of Gauss.

Let $0 < a < b$. We define two sequences a_n and b_n by

$$a_0 = a, \quad b_0 = b, \quad a_{n+1} = \sqrt{a_n b_n}, \quad b_{n+1} = \frac{a_n + b_n}{2}$$

a) Show that a_n is monotone increasing, b_n is monotone decreasing, and $a_n < b_n$.

b) Show that $b_{n+1} - a_{n+1} = \frac{(b_n - a_n)^2}{8b_{n+2}}$

c) Show that $\lim_{n \to \infty} a_n = \lim_{n \to \infty} b_n = g$

d) Write a program which prints the sequences a_n and b_n with initial values $a_0 = \frac{1}{2}$, $b_0 = 1$. Note the rapid convergence to the same limit (quadratic convergence).

e) If we define $a_0 = a$, $b_0 = b$, $a_{n+1} = \sqrt{a_n b_{n+1}}$, $b_{n+1} = \frac{a_n + b_n}{2}$

the speed of convergence is quite different.

Show that $a_n^2 - b_n^2 = \frac{a_0^2 - b_0^2}{4^n}$, and $a_n - b_n = \frac{1}{4^n} \frac{a_0^2 - b_0^2}{2a_{n+1}}$

(linear convergence). Determine $\lim_{n \to \infty} a_n = \lim_{n \to \infty} b_n = g$.

for $a_0 = \frac{1}{2}$, $b_0 = 1$. (See 3.4)

18. Show that the series $x_{n+1} = x_n (2 - ax_n)$, $a > 0$ converges quadratically to $\frac{1}{a}$ for suitable initial values x_0.

<u>Hint</u>. Write $x_n = \frac{1}{a}(1 - \varepsilon_n)$.

19. <u>Exponential towers.</u> Fig. 4.19a shows a program which, when a real number $A > 0$ is input, prints an infinite sequence. By experimentation with the computer, find the maximum value of A (approximately) for which the sequence converges.

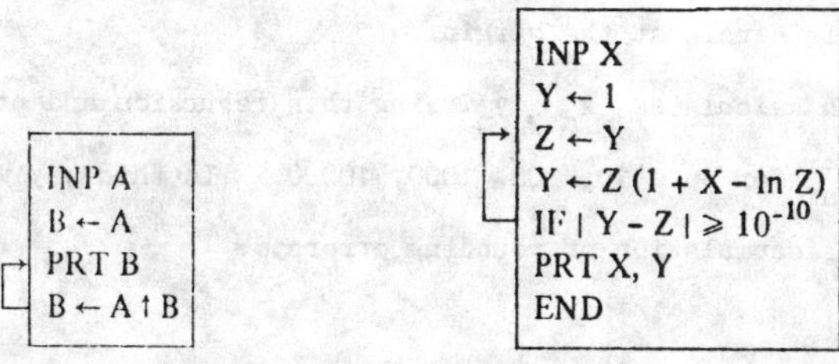

Fig. 4.19a

Fig. 4.19b

20. Calculate π as the solution of the equation $\sin x = 0$ by Newton's method. Start with $x_0 = 3, 2, 4, 1{\cdot}6, 4{\cdot}5$.

21. <u>The inverse function via Newton's method</u>

Using Newton's method one can construct an algorithm for the inverse f^{-1} of a monotone function f. Let the function $y = \ln x$ be available on a computer. We seek a program which when x is input, will output the solution y of the equation $x = \ln y$. Show that Fig. 4.19b performs this. By experimentation with different input values ascertain the values of x for which the algorithm outputs $y = e^x$.

22. Let $a > 0$. Both $f(x) = x^3 - a$ and $h(x) = x^2 - \frac{a}{x}$ have the zero $\sqrt[3]{a}$. Use equation (7) to give two iteration sequences which converge to $\sqrt[3]{a}$. Verify that for $a = 8$, $a = 2$, $a = 10$, h gives a more rapidly convergent sequence (cubic convergence).

23. Solve the equation $x^{20} - 1 = 0$ using Newton's method, with initial value $x_0 = \frac{1}{2}$. Note the very slow initial convergence (with factor $\frac{19}{20}$). Quadratic convergence occurs only in the immediate neighbourhood of $x = 1$.

24. Putting the computer to the test

We define a sequence of points (x_n, y_n) by the recursion

$$x_o = 1, \quad y_o = 0, \quad x_{n+1} = \frac{3x_n - 4y_n}{5}, \quad y_{n+1} = \frac{4x_n + 3y_n}{5}.$$

Since $x_{n+1}^2 + y_{n+1}^2 = x_n^2 + y_n^2 = x_o^2 + y_o^2 = 1$, all points of the sequence lie on a unit circle at the origin.

Write a program which calculates x_n, y_n using this recursion and prints the value of $x_n^2 + y_n^2$ for $n = 10, 100, 1000, 10000$. In this way we can easily study the accumulation of rounding errors.

4.1 The maximum of a unimodal function

We construct an interesting program for determining the maximum of a function. For this we restrict ourselves to the so called unimodal functions. F is unimodal on $[a, b]$ if it has exactly one (local) maximum. Fig. 4.20 shows four unimodal curves. A unimodal function need not be either differentiable or continuous.

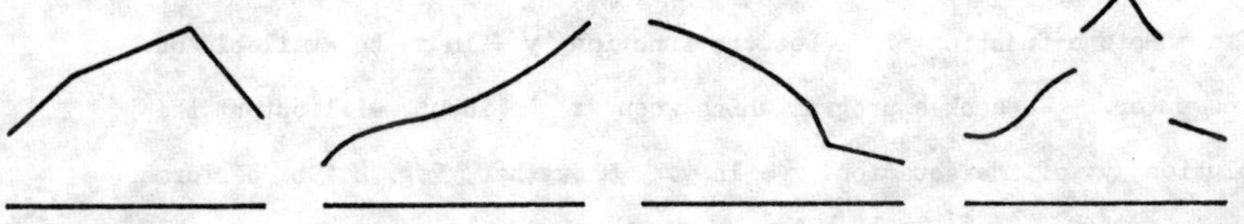

Fig. 4.20

Halving (bisection) method. The kernel of this program consists of the process of halving the interval in which the maximum lies. The initial interval M - H, M + H with mid-point M is divided into four equal parts. (Fig. 4.21). The maximum lies in one of the following intervals of length H:

$$I_1 = [M - H, M], \quad I_2 = \left[M - \frac{H}{2}, M + \frac{H}{2}\right], \quad I_3 = [M, M + H]$$

If $F(M - \frac{H}{2}) > F(M)$, then the maximum lies in I_1.

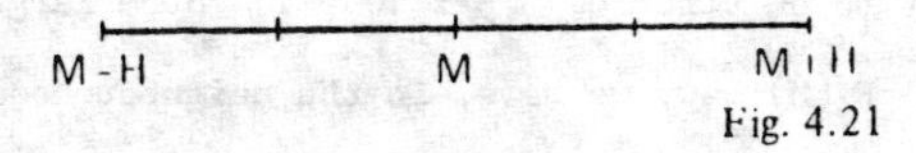

M-H M M+H

Fig. 4.21

If $F(M + \frac{H}{2}) > F(M)$ then the maximum lies in I_3.

If neither of these inequalities holds, the maximum lies in I_2. (Fig. 4.22)

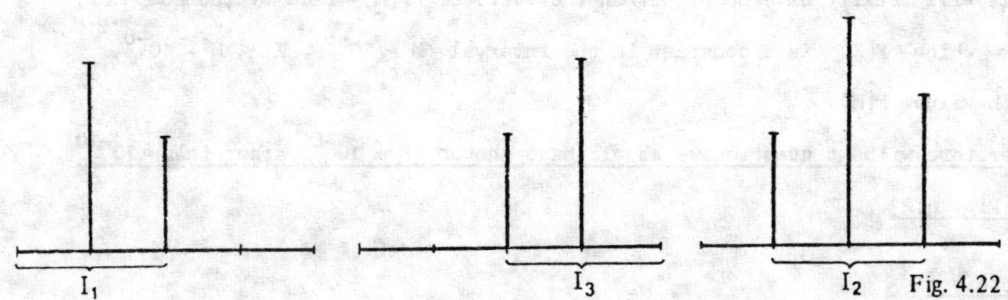

Fig. 4.22

The program in Fig. 4.23 is short, elegant and instructive. It chooses whichever one of the intervals I_1, I_2, I_3 contains the maximum, replaces M by the mid point of the chosen interval and halves H. The new H is compared with the allowable error E. If $H \leq E$, then M is the desired approximation for the location of the maximum.

```
 10 READ M, H, E
 20 DEF FNF (X) = 4*X↑4/(1 + X↑6)
 30 A = FNF (M)
 40 IF FNF (M - H/2) > A THEN M = M - H/2
 50 IF FNF (M + H/2) > A THEN M = M + H/2
 60 H = H/2
 70 IF H > E THEN 30
 80 PRINT M, FNF (M)
 90 DATA 1, 1, 1E - 10
100 END
    1.122462048    2.116534736
```

Fig. 4.23

The exact values are $M = \sqrt[6]{2} = 1.122467041$; $F(M) = \frac{4}{3}\sqrt[6]{16} = 2.116534736$.

The program gives the maximal value with 10 correct digits, whereas the location of the maximum is subject to an error of 5×10^{-6}. Such large errors cannot be avoided. Since $F'(M) = 0$ we have, in the neighbourhood of M,

$$F(M + H) = F(M) + \frac{F''(\bar{M})H^2}{2} \quad (\text{where } M < \bar{M} < M + H).$$

An increment of 10^{-6} in M produces one of order 10^{-12} in F(M) and this will usually have no effect on the value of F(M) held by the computer, for which F(X) is a constant in the interval $M - 10^{-6} < X < M + 10^{-6}$, with value F(M).

Therefore without question we should have chosen $E = 10^{-6}$ rather than 10^{-10} in Fig. 4.23.

Exercises

1. We describe a simple method of determining the maximum of a function f in $[a, b]$. We start at a and move to the right in steps of length h, until for the first time, f grows smaller. Then we reduce the step length 10 fold, turn around and step to the left, until f begins to decrease. Then we turn around, reduce the step length 10 fold, and so on. This process is continued until the step lengths become less than some pre-assigned error bound ε. Translate this description into a BASIC program and determine the maximum value of $f(x) = 4x - \frac{x^4}{4}$ for $x > 0$. Compare with the exact value, $x = \sqrt[3]{4}$, for the location of the maximum.

   Remark If one wishes only to solve a problem, the computation time is immaterial. In such a case one would prefer this wasteful method to the halving method, since it is easier to program. The method in exercise 5 is better still.

2. Modify the program in exercise 1 so that it locates a minimum value. Determine, for $x > 0$, the location of the minimum values of

   a) $f(x) = x^2 + \frac{8}{x}$, b) $f(x) = x^x$.

3. We describe a maximum algorithm which is related to that in exercise 1. The function of f has a maximum in $[a, b]$

1. Start at a, choose a step length h and an error bound ε.

2. Move to the right in steps of h, until f begins to decrease.

3. Go back one step and print x.

4. Put $h \leftarrow \frac{h}{10}$. If $h \geqslant \varepsilon$ go back to step 2. Otherwise HALT.

This algorithm contains an error, through which it sometimes breaks down. Explain the error, using a diagram, and correct it.

4. Determine the maximum of a) $y = e^{x} \sin x$ in $[0, \pi]$

b) $y = xe^{-x}$ for $x > 0$.

5. We seek the maximum of the unimodal function f in the interval $[a, b]$. We put $h = b - a$ and choose an error bound ε.
If $f(x + \frac{h}{3}) < f(b - \frac{h}{3})$ we can eliminate the left hand third of the interval, otherwise we eliminate the right hand third.
This step is repeated until $h < \varepsilon$. Then we take $\frac{a + b}{2}$ as the location of the maximum. Write a program for this.

4.3 Numerical integration

4.3.1 Trapezium rule, mid-point rule, and Simpson's rule

The shaded region in Fig. 4.24 has area

(1) $$I = \int_a^b f(x)\, dx.$$

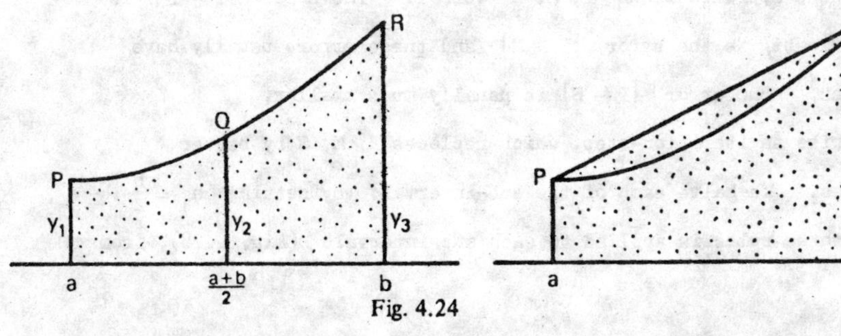

Fig. 4.24 Fig. 4.25

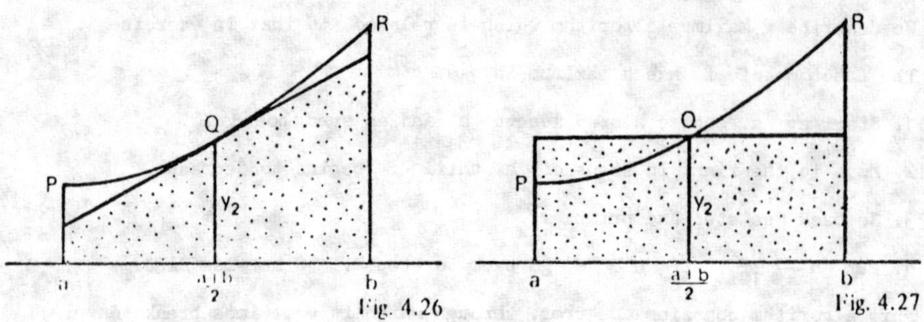

Fig. 4.26 Fig. 4.27

Often the integral function of f is not known or not tabulated. Then I must be evaluated using a computer. There are three well-known methods of approximating I, in which f is replaced by

a) the chord PR (Fig. 4.25);

b) the tangent at Q (Fig. 4.26), or, with the same result, the horizontal line through Q. (Fig. 4.27)

c) a quadratic curve, $y = px^2 + qx + r$ through P, Q, R.

The three methods are called Trapezium rule, Mid-point rule and Simpson's rule.

If we write $h = b - a$, the Trapezium rule is

(2) $$T = \frac{h}{2}(y_1 + y_3)$$

The Mid-point rule is

(3) $$M = hy_2$$

As will be shown later, the Simpson's rule is

(4) $$S = \frac{h}{6}(y_1 + 4y_2 + y_3) = \frac{T + 2M}{3}$$

So we have three approximations T, M, S for I. The error $|I - T|$ is about twice as big as the error $|I - M|$ and these errors usually have opposite signs. The error $|I - S|$ is usually much smaller.

Next we describe an iteration-step, which replaces T, M, S by better approximations. We halve each of the sub-intervals by setting $h_1 = \frac{h}{2}$. Each of the three rules is applied to each sub-intervals (Fig. 4.28), from which

$$T_1 = \frac{h}{2}\cdot\frac{y_1 + y_2}{2} + \frac{h}{2}\cdot\frac{y_2 + y_3}{2} = \frac{y_1 + y_3}{2}\cdot\frac{h}{2} + y_2\frac{h}{2} = \frac{T + M}{2},$$

$$M_1 = \frac{h}{2}\, y_4 + \frac{h}{2}\, y_5 = (y_4 + y_5)\, h_1$$

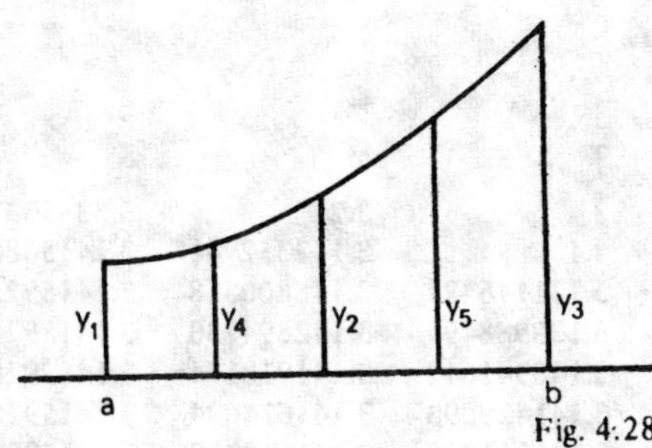

Fig. 4.28

```
FOR X = A + H/2 TO B STEP H
M = M + FNF (X)
NEXT X
M = M*H
```

Fig. 4.29

To obtain M_1 we add the ordinates in the newly-created sub intervals and multiply the sum by h_1. We achieve this in BASIC through the four lines in Fig. 4.29 where H is written instead of H_1.

As soon as we have T_1 and M_1 we calculate

$$S_1 = \frac{T_1 + 2M_1}{3}$$

The program in Fig. 4.30 was written for $\int_0^1 \frac{4}{1 + x^2}\, dx = \pi$

If we wish to evaluate a different integral, we have only to alter the lines 10 and 170. 10 defines the function f, 20 reads the limits of integration A and B and the allowable error E. 30 prints the table heading. In 50 the division by 2 is omitted. This is remedied in lines 60 and 70. 70 to 160 carry out the iteration. We shall explain line 160 in more detail. $|T - M|$ usually overestimates the absolute errors $|I - T|$ and $|I - M|$ while $|S|$ is very close to $|I|$. That is, $\frac{|T - M|}{|S|}$ is an overestimate of the relative errors of M and T. We break off when this error is less than 10^{-6}. The relative error in S will be much smaller (of order of magnitude 10^{-12}). If a computer is available, the program in Fig. 4.30 is sufficient for all practical purposes. If one has only a pocket calculator, then a better procedure is preferable.

```
10 DEF FNF (X) = 4/(1 + X↑2)
20 READ A, B, E
30 PRINT "T", "M", "S"
40 H = B - A
50 T = (FNF (A) + FNF (B) )*H
60 M = 0
70      T = (T + M)/2
80      M = 0
90      FOR X = A + H/2 TO B STEP H
100          M = M + FNF (X)
110     NEXT X
120     M = M*H
130     S = (T + 2*M)/3
140     PRINT T, M, S
150     H = H/2
160 IF ABS (T - M)/ABS (S) > E THEN 70
170 DATA 0, 1, 1E - 6
180 END
```

T	M	S
3	3.2	3.333333333
3.1	3.162352941	3.141568627
3.131176471	3.146800518	3.141592502
3.138988494	3.142894730	3.141592651
3.140941612	3.141918174	3.141592654
3.141429893	3.141674034	3.141592654
3.141551963	3.141612999	3.141592654
3.141582481	3.141597740	3.141592654
3.141590110	3.141593925	3.141592654
3.141592018	3.141592971	3.141592654

Fig. 4.30

4.3.2 The Romberg procedure

The Trapezium rule gives a poor approximation for the integral I in (1)

$$T_0 = \frac{h}{2}(f(a) + f(b)).$$

Using the foregoing halving process in the interval $[a, b]$ and applying the Trapezium rule in each sub-interval, we obtain a sequence converging to I.

(5) $T_0, T_1, T_2, T_3, \ldots$

It can be shown that $T_n = I + a_1 4^{-n} + a_2 4^{-2n} + a_3 4^{-3n} \ldots$

where the a_i are independent of n. This was proved in 3.6 for the circle and the hyperbola.

So we can set out a Romberg table as in 3.6:

$$\begin{array}{lllll}
T_0 & & & & \\
 & T_0' & & & \\
T_1 & & T_0'' & & \\
 & T_1' & & T_0''' & \\
T_2 & & T_1'' & & T_0^{(4)} \\
 & T_2' & & T_1''' & \\
T_3 & & T_2'' & & \vdots \\
 & T_3' & & \vdots & \\
T_4 & & \vdots & & \\
\vdots & \vdots & & &
\end{array}
\qquad
\begin{aligned}
T_n' &= T_{n+1} + \frac{T_{n+1} - T_n}{3} \\
T_n'' &= T_{n+1}' + \frac{T_{n+1}' - T_n'}{15} \\
T_n''' &= T_{n+1}'' + \frac{T_{n+1}'' - T_n''}{63} \\
T_n^{(4)} &= T_{n+1}''' + \frac{T_{n+1}''' - T_n'''}{255} \\
\vdots \quad &\quad \vdots \qquad \quad \vdots
\end{aligned}$$

The first column has convergence factor $\frac{1}{4}$. Each successive column converges four times as fast as its predecessor. The diagonal $T_0, T_0', T_0'', T_0''', \ldots$ converges very rapidly. This so-called Romberg procedure is one of the best methods of numerical integration.

Example We calculate the integral

$$I = \int_0^1 \frac{\sin x}{x}\, dx$$

We have $T_0 = \frac{1 + \sin 1}{2}$, $M_0 = 2 \sin \frac{1}{2}$, $T_1 = \frac{T_0 + M_0}{2}$,

$$M_1 = 2 \sin \tfrac{1}{4} + \tfrac{2}{3} \sin \tfrac{3}{4},$$

$$T_2 = \frac{T_1 + M_1}{2}, \quad M_2 = 2 \sin \tfrac{1}{8} + \tfrac{2}{3} \sin \tfrac{3}{8} + \tfrac{2}{5} \sin \tfrac{5}{8} + \tfrac{2}{7} \sin \tfrac{7}{8},$$

$$T_3 = \frac{T_2 + M_2}{2}$$

from which follows the Romberg table:

0·9207354924			
	0·9461458823		
0·9397932848		0·9461458823	
	0·9460869339		0·9460830704
0·9445135217		0·9460830694	
	0·9460833109		
0·9456908636			

Thus $I = 0{\cdot}9460830704$. The correct value is $I = 0{\cdot}94608307036\ldots$ The Romberg table should have at most 7 columns, otherwise the rounding errors become too large. There is no such limit on the number of lines.

Exercises

1. Determine by the Romberg procedure

a) $\displaystyle\int_0^2 xe^{-x}\,dx$ b) $\displaystyle\int_1^2 \frac{e^x}{x}\,dx$ c) $\displaystyle\int_0^1 e^{-x^2}\,dx$

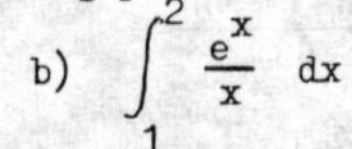

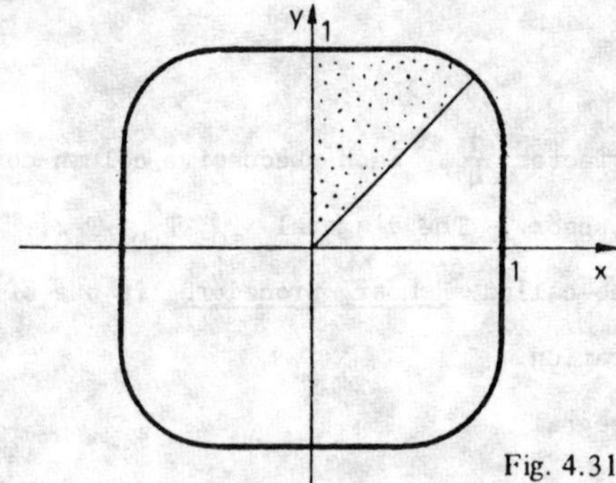

Fig. 4.31

2. Fig. 4.31 shows the "super circle" $x^4 + y^4 = 1$. We wish to calculate its area in two ways. Show that

$$I = 4 \text{ x the first quadrant} = 4\int_0^1 \sqrt[4]{1 - x^4}\,dx$$

$$\text{and } I = 8 \text{ x the shaded octant} = 8\int_0^{\sqrt[4]{0.5}} \sqrt[4]{1 - x^4}\,dx - 2\sqrt{2}$$

Estimate I by both methods using the program in Fig. 4.30.

3. Let $\pi(n)$ be the number of primes $\leq n$. The Prime Number theorem states

$$\pi(n) \sim \int_2^n \frac{dx}{\ln x}$$

We wish to calculate the integral accurately for $n = 10^9$ and compare the result with $\pi(10^9) = 50\,847\,534$. Fig. 4.32 shows the required region. Using the mid-point rule with 1, 2, 4, 8, . . ., 2^{12} subintervals, we obtain the approximations in table 4.33.

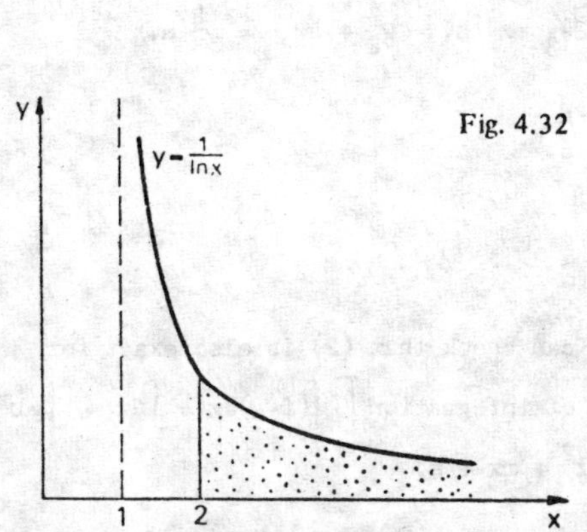

Fig. 4.32

n	M_n
0	49 924 816.48
1	50 324 328.92
2	50 558 006.71
3	50 689 175.19
4	50 761 465.98
5	50 801 043.43
6	50 822 688.75
7	50 834 547.40
8	50 841 065.33
9	50 844 662.74
10	50 846 657.95
11	50 847 770.69
12	50 848 395.14

Fig. 4.33

a) Write a program which prints this table given the input

$A = 2, \quad B = 10^9$.

b) The sequence M_0 . . . , M_{12} does not have convergence factor $\frac{1}{4}$. This is a consequence of the very large interval lengths. Even for $n = 12$ each of the 2^{12} trapezia has width $\doteq$ 250000. Here the region must be further subdivided (e.g. from 2 to 10^3, from 10^3 to 10^7, and from 10^7 to 10^9) and the sub areas calculated separately. Then the convergence factor of $\frac{1}{4}$ begins to show itself. Using the Romberg procedure on M_{10}, M_{11}, M_{12}, we obtain

$$\int_2^{10^9} \frac{dx}{\ln x} \doteq 50\,849\,235$$

Check this! (The trapezium rule gives much poorer results)

4.3.3. Proof of the Simpson rule and the Hermite rule

a) In Fig. 4.34 we seek a formula

$$(1) \qquad \int_0^h f(x)dx = w_1y_1 + w_2y_2 + w_3y_3 ,$$

which is exact for $f(x) = 1, x, x^2$. Using these functions in turn,

we have $w_1 + w_2 + w_3 = h, \quad w_2 + 2w_3 = h, \quad w_2 + 4w_3 = \frac{4}{3}h.$

so that $w_1 = w_3 = \frac{h}{6}, \quad w_2 = \frac{4}{6}h.$

Thus

$$(2) \qquad S = \frac{h}{6}(y_1 + 4y_2 + y_3)$$

is exact for $f(x) = 1, x, x^2$. One can check that (2) is also exact for x^3. Because of the linear property of integration $\left[\int(f+g)dx = \int fdx + \int gdx\right]$ it is also exact for $f(x) = px^3 + qx^2 + rx + s$.

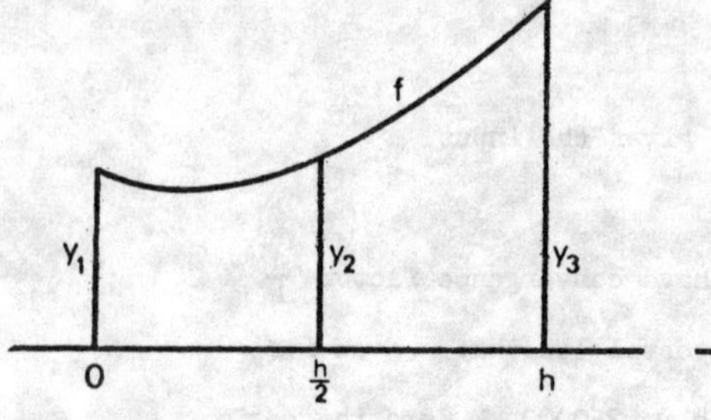

Fig. 4.34

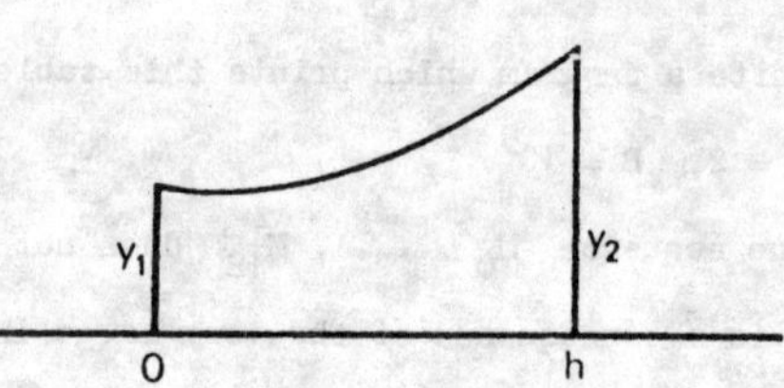

Fig. 4.35

b) In Fig. 4.35 we seek a formula

$$(3) \qquad \int_0^h f(x)\,dx = v_1y_1 + v_2y_2 + w_1y'_1 + w_2y'_2$$

which is exact for $f = 1, x, x^2, x^3$. Using these functions in turn, we have

$$v_1 + v_2 = h, \quad hv_2 + w_1 + w_2 = \frac{h^2}{2}, \quad hv_2 + 2w_2 = \frac{h^2}{3}, \quad hv_2 + 3w_2 = \frac{h^2}{4},$$

with the solutions

$$v_1 = v_2 = \frac{h}{2}, \quad w_1 = -w_2 = \frac{h^2}{12}$$

Thus we have Hermite's rule

$$(4) \qquad \int_a^b f(x)dx = \frac{h}{2}\Big[f(a) + f(b)\Big] + \frac{h^2}{12}\Big[f'(a) - f'(b)\Big]$$

which is exact for the cubic curve $f(x) = px^3 + qx^2 + rx + s$.

It has about the same accuracy as the Simpson rule and is particularly useful when the derivative $f'(x)$ is easily found.

Exercises

1. Let $T = \frac{h}{2}\Big[f(a) + f(b)\Big]$, $V = \frac{h}{2}\Big[f(a) + f(b)\Big] + \frac{h^2}{12}\Big[f'(a) - f'(b)\Big]$

a) Show that the shaded region in Fig. 4.36 has area

$$U = \frac{h}{2}\Big[f(a) + f(b)\Big] + \frac{h^2}{8}\Big[f'(a) - f'(b)\Big]$$

b) Show that $V = \frac{T + 2U}{3}$

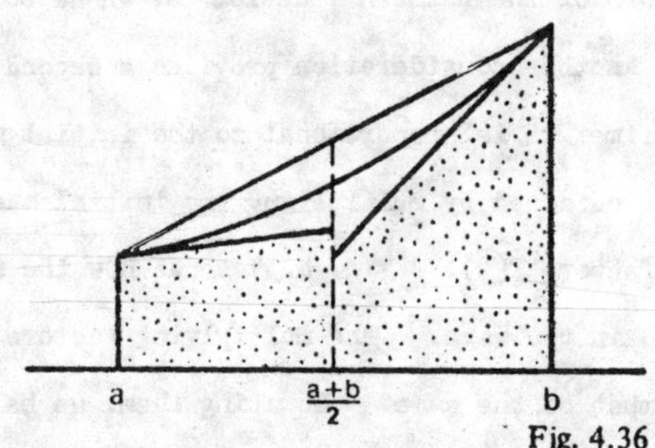

Fig. 4.36

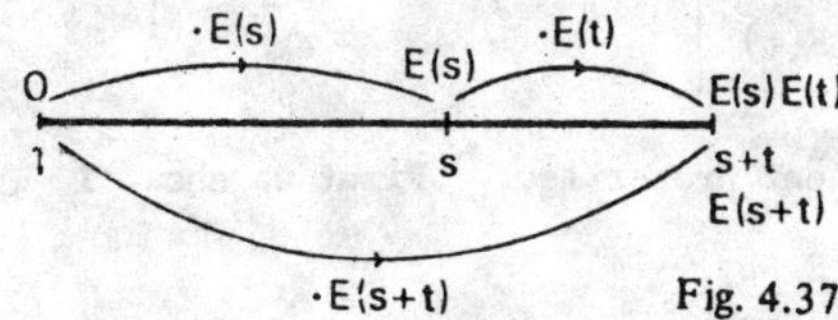

Fig. 4.37

4.4 Differential equations

4.4.1 Growth, Decay, and oscillation

Important natural processes lead in a direct way to two simple differential equations which are fertile ground for numerical methods.

a) The exponential function

A typical growth process is the growth of a colony of bacteria in the absence of competition for space and food. A typical decay process is that of radioactive decay. We begin at time $t = 0$ with initial mass 1. At time t let the mass of the substance be $E(t)$. In both processes $E'(t)$ is proportional to $E(t)$. i.e.

$$\text{(I)} \qquad E'(t) = a.\,E(t), \qquad E(0) = 1. \qquad \begin{array}{l} a > 0 \text{ growth} \\ a < 0 \text{ decay} \end{array}$$

This is the differential equation for the unknown function E, whose solution will occupy us further below. Another consideration provides a second property of E. The mass at time t is proportional to the initial mass. That is, the mass at time t is obtained by multiplying the initial mass by the respective growth or decay factor $E(t)$. Fig. 4.37 shows how the mass at time $s + t$ can be expressed in two ways. The multiplying factors on both sets of transition arrows must be the same. Equating them, we have

$$\text{(II)} \qquad \boxed{E\,(s + t) = E(s)\,.\,E(t)}$$

We show that I and II are equivalent properties. First we show I $\Rightarrow$ II. For this we define the function F by

$$\text{(1)} \qquad F(t) = E(h-t)\,E(t)$$

Differentiation with respect to t gives, by I,

$$F'(t) = -aE\,(h - t).E(t) + E(h - t).a.E(t) = 0$$

Thus F is independent of t, or

$$F(t) = F(0) = E(h)$$

or

(2) $$E(h - t)\, E(t) = E(h)$$

Putting $h = 0$ gives

(3) $$E(t)\, E(-t) = 1$$

and $h = s + t$ gives

(4) $$E(s + t) = E(s)\, E(t)$$

So I => II has been proved.

It follows from (3) that $E(t)$ is never zero and therefore since $E(0) = 1$, $E(t)$ is always positive.

Now we show II => I.

From II, by differentiation with respect to t,

$$E'(s + t) = E(s).\, E'(t).$$

and putting $t = 0$,

$$E'(s) = E'(0)\, E(s).$$

If we write $E'(0) = a$, this becomes

$$E'(s) = aE(s).$$

From II, with $t = 0$, we have $E(0) = 1$, so that II => I has been proved.

From II with $s = t$ we have

$$E(2t) = E(t)^2.$$

By mathematical induction it follows that

(5) $$E(nt) = E(t)^n$$

If t is rational, $t = \frac{m}{n}$, then $nt = m.1$ and (5) gives

$$E(nt) = E(m.1) \Rightarrow E(t)^n = E(1)^m \Rightarrow E(t) = E(1)^{\frac{m}{n}}$$

That is

(P) $$\boxed{E\left(\tfrac{m}{n}\right) = b^{\frac{m}{n}}, \quad b = E(1) > 0.}$$

The differential equation I can be solved iteratively in two ways.

1. <u>The crude method of Euler polygons</u>

We put $x_k = \frac{kt}{n}$, $E(x_k) = y_k$.

$E'(x_k)$ is approximated by the difference quotient

$$\frac{y_{k+1} - y_k}{x_{k+1} - x_k}$$

This gives (Fig. 4.38)

$$y_0 = 1, \quad y_{k+1} = \left(1 + \frac{at}{n}\right) y_k$$

from which it follows that

$$y_n = \left(1 + \frac{at}{n}\right)^n$$

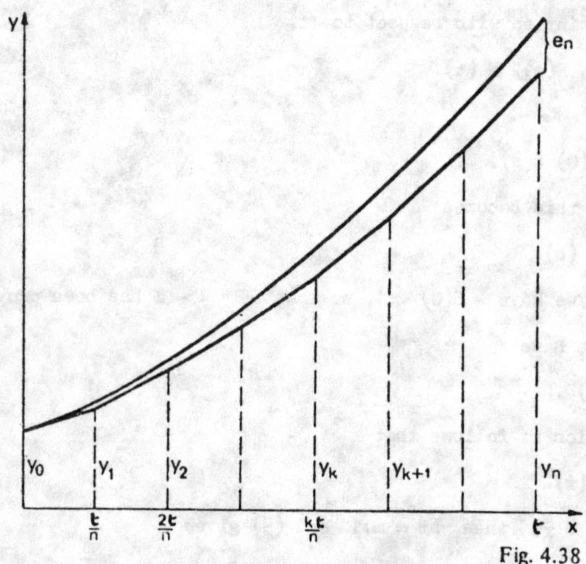

Fig. 4.38

This is an estimate of $E(t)$. At the point t the error is $e_n(t) = E(t) - y_n$.

We conjecture that $e_n(t) \to 0$ for $n \to \infty$. i.e. that

(III) $$E(t) = \lim_{n \to \infty} \left(1 + \frac{at}{n}\right)^n$$

When $a = 1$, E is called the <u>exponential function</u> and is denoted by <u>exp</u>.

Thus

(E) $$\exp(t) = \lim_{n \to \infty} \left(1 + \frac{t}{n}\right)^n$$

2. Picard iteration

The differential equation I is equivalent to the integral equation

$$\text{(IV)} \qquad \boxed{E(t) = 1 + a\int_0^t E(x)dx}$$

In fact, IV is obtained by integrating I between the limits 0 and t.

IV has the form

$$E = \Phi(E)$$

which invites the iteration $E_{n+1} = \Phi(E_n)$. i.e.

$$E_{n+1}(t) = 1 + a\int_0^t E_n(x)dx.$$

We start with the first approximation $E_0(t) = 1$ and obtain the series

$$E_0(t) = 1$$

$$E_1(t) = 1 + at$$

$$E_2(t) = 1 + at + \frac{a^2t^2}{2!}$$

$$\dots \quad \dots \quad \dots$$

$$E_n(t) = \sum_{i=0}^{n} \frac{(at)^i}{i!}$$

So we conjecture that

$$\text{(V)} \qquad \boxed{E(t) = \lim_{n\to\infty} E_n(t) = \sum_{n\geqslant 0} \frac{(at)^n}{n!}}$$

In particular

$$\text{(VI)} \qquad \boxed{\exp(t) = \sum_{n\geqslant 0} \frac{t^n}{n!}}$$

The number e is defined as exp(t). We conjecture therefore

$$\text{(VII)} \qquad \boxed{e = \lim_{n\to\infty} (1 + \frac{1}{n})^n = \sum_{n\geqslant 0} \frac{1}{n!}}$$

and

(VIII) $$e^{-1} = \frac{1}{e} = \lim_{n \to \infty} (1 - \frac{1}{n})^n = \sum_{n \geq 0} \frac{(-1)^n}{n!}$$

In the following exercises, e and e^{-1} are to be calculated using VII and VIII.

Exercises

1. Let $e_n = \sum_{i=0}^{n} \frac{1}{i!} = 1 + 1 + \frac{1}{2!} + \frac{1}{3!} + \ldots + \frac{1}{n!}$ (Note that $0! = 1! = 1$, $n! = 1, 2, 3, 4..n$)

a) Write a program which prints e_n for $n = 0, 1, 2, \ldots 15$.

b) Write a program to calculate e_n based on the following transformation

$$e_n = 1 + \frac{1}{1}(1 + \frac{1}{2}(1 + \frac{1}{3}(1 + \frac{1}{4}(1 + \ldots + \frac{1}{n-1}(1 + \frac{1}{n}) \ldots))))$$

c) Show by suitable approximation by a geometric series, that

$$e - e_n = \frac{1}{(n+1)!} + \frac{1}{(n+2)!} + \ldots < \frac{n+2}{n+1} \quad \frac{1}{(n+1)!}$$

Use this to estimate $e - e_{15}$.

2. Let $c_n = \sum_{i=0}^{n} \frac{(-1)^i}{i!} = 1 - 1 + \frac{1}{2!} - \frac{1}{3!} + \frac{1}{4!} - \ldots + \frac{(-1)^n}{n!}$

Write a program which prints c_n for $n = 0, 1, 2, \ldots, 15$.

3. Let $a_n = (1 + \frac{1}{n})^n$ and $b_n = (1 - \frac{1}{n})^n$. Using the rapid exponentiation programs in Fig. 1.30 or Fig. 2.5b, calculate a_n and b_n for $n = 10, 10^2, \ldots 10^{12}$.

4. Let f be a given error bound. Write a program which for input x, f will output exp(x). Use VI to calculate exp(x). If s is the current sum and t the next member of the series, terminate the program when $\left|\frac{t}{s}\right| < f$.

The function exp is defined by exp = exp, exp(0) = 1, and we write exp(1) = e. On account of (P), we also write

$$\exp x = e^x$$

From $\exp' = \exp$, it follows that

$$\int_a^b e^x dx = e^b - e^a.$$

The sequences e_n and c_n in exercises 1 and 2 give their limiting values to 10 significant figures without much labour.

$$e = 2{\cdot}71828\ 18284\ 59045\ldots\ ,\quad e^{-1} = 0{\cdot}36787\ 94411\ 71442\ \ldots$$

We consider now further sequences which converge to e at different rates, and we investigate these numerically.

1. Example

The sequences a_n and b_n in exercise 3 are not suitable for precise calculation of e and e^{-1} respectively. For large n there is a catastrophic loss of accuracy due to rounding. So we shall try to obtain the limit value e by extrapolation through the Romberg table. Unfortunately a_n and b_n arise from approximations to an area using rectangles rather than trapezia. The curved "trapezium" in Fig. 4.39 has area

$$e^{\frac{1}{n}} - 1.$$

The small and large rectangles have areas

$$\frac{1}{n} \text{ and } \frac{1}{n}\, e^{\frac{1}{n}}$$

From

$$\frac{1}{n} < e^{\frac{1}{n}} - 1 < \frac{1}{n}\, e^{\frac{1}{n}} \qquad \text{we quickly obtain}$$

$$(1 + \frac{1}{n})^n < e < (1 - \frac{1}{n})^{-n}$$

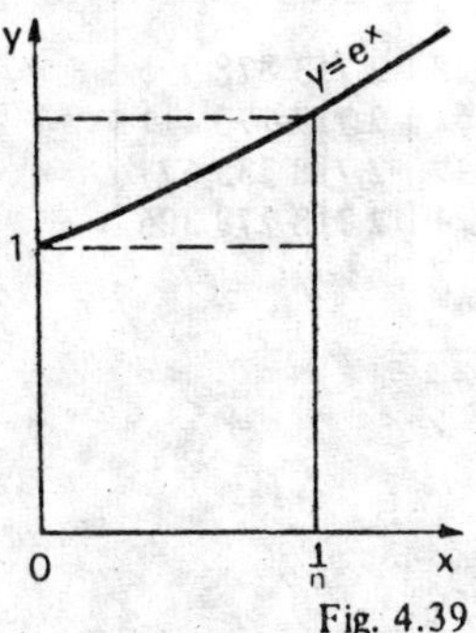

Fig. 4.39

When n is doubled the error in a trapezium approximation becomes about 4 times as small. In a rectangle approximation it is approximately halved, as one can see with the help of a diagram. We write

$$A_n = a_{2^n} = \left(1 + \frac{1}{2^n} \right)^{2^n}$$

It can be shown that

$$A_n = e + c_1 2^{-n} + c_2 2^{-2n} + c_3 2^{-3n} + \ldots$$

From the sequence A_n with convergence factor $\frac{1}{2}$ we construct sequences A'_n, A''_n, A'''_n,.. with convergence factors $\frac{1}{4}$, $\frac{1}{8}$, $\frac{1}{16}$, . . . :

$$A'_n = A_{n+1} + \frac{1}{2^1 - 1}(A_{n+1} - A_n) = e + c_1 2^{-2n} + c_2 2^{-3n} + \ldots$$

$$A''_n = A'_{n+1} + \frac{1}{2^2 - 1}(A'_{n+1} - A'_n) = e + c_1 2^{-3n} + c_2 2^{-4n} + \ldots$$

$$A'''_n = A''_{n+1} + \frac{1}{2^3 - 1}(A''_{n+1} - A'_n) = e + c_1 2^{-4n} + c_1 2^{-5n} + \ldots$$

$$\vdots \qquad \vdots \qquad \vdots \qquad\qquad \vdots \qquad \vdots \qquad \vdots \qquad \vdots$$

Table 4.40 shows the result of the extrapolation.

The best approximation 2·718281729 has an absolute error of 10^{-7}.

n	A_n	A'_n	A''_n	A'''_n	$A_n^{(4)}$	$A_n^{(5)}$	$A_n^{(6)}$
0	2.000 000 000						
1	2.250 000 000	2.500 000 000					
2	2.441 406 250	2.632 812 500	2.677 083 334				
3	2.565 784 514	2.690 162 778	2.709 279 538	2.713 878 995			
4	2.637 928 499	2.710 072 483	2.716 709 051	2.717 770 411	2.718 029 838		
5	2.676 990 128	2.716 051 757	2.718 044 849	2.718 235 677	2.718 266 694	2.718 274 335	
6	2.697 344 955	2.717 699 782	2.718 249 124	2.718 278 306	2.718 281 148	2.718 281 614	2.718 281 729

Fig. 4.40

2. Example

Fig. 4.41 shows the graph of $y = e^x$ from 0 to $\frac{1}{n}$.

Curve-, chord- and tangent-trapezia have the areas

$$e^{\frac{1}{n}} - e^0 , \quad \frac{1}{2n}(1 + e^{\frac{1}{n}}) , \quad \frac{1}{2n}(2 + \frac{1}{n}) \quad \text{respectively.}$$

A short calculation gives

$$1 + \frac{1}{n} + \frac{1}{2n^2} < e^{\frac{1}{n}} < 1 + \frac{1}{n - 0{\cdot}5}$$

(IX) $$(1 + \frac{1}{n} + \frac{1}{2n^2})^n < e < (1 + \frac{1}{n - 0{\cdot}5})^n$$

Exercise 5

Write a program which gives an interval approximation for e, using IX, for $n = 10, 10^2, 10^3, 10^4$. Use the standard function ↑ .

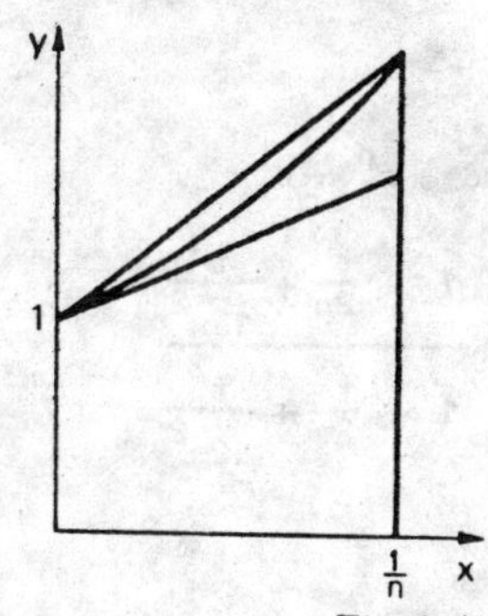

Fig. 4.41

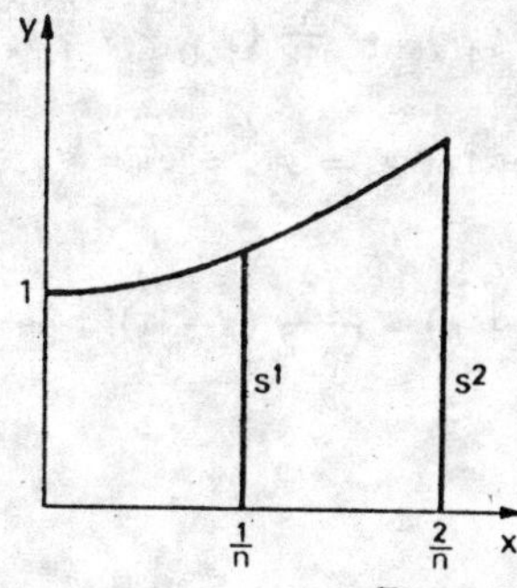

Fig. 4.42

3. Example

Fig. 4.42 shows the graph of exp from 0 to $\frac{2}{n}$. We write $e^{\frac{1}{n}} = s$, i.e. $e^{\frac{2}{n}} = s^2$. The area under the curve is $s^2 - 1$. Simpson's rule gives the approximation

$$\frac{1}{3n} (1 + 4s + s^2)$$

From

$$s^2 - 1 \doteq \frac{1}{3n}(1 + 4s + s^2)$$

we have

$$s \doteq \frac{2 + \sqrt{9n^2 + 3}}{3n - 1} , \qquad e = s^n \doteq \left(\frac{2 + \sqrt{9n^2 + 3}}{3n - 1} \right)^n .$$

We may therefore expect that the sequence

$$e_n = \left(\frac{2 + \sqrt{9n^2 + 3}}{3n - 1} \right)^n$$

converges very quickly to e.

Exercise 6

Print a table of e_n for $n = 1, 2, 2^2, \ldots 2^7$. The table is easily produced with a pocket calculator.

4. Example

Fig. 4.43 shows the graph of exp from 0 to $\frac{t}{n}$. The area under the curve is $e^{\frac{t}{n}} - 1$. The Hermite rule gives the approximation

$$V = \frac{h}{2} (y_0 + y_1) + \frac{h^2}{12} (y'_0 - y'_1).$$

With $h = \frac{t}{n}$, $y_0 = y'_0 = 1$, $y_1 = y'_1 = e^{\frac{t}{n}} = s$ and $s^n = e^t$ we have

$$s - 1 \doteq \frac{t}{2n} (1 + s) - \frac{t^2}{12n^2} (s - 1) , \quad s \doteq \frac{1 + \frac{t}{2n} + \frac{t^2}{12n^2}}{1 - \frac{t}{2n} + \frac{t^2}{12n^2}}$$

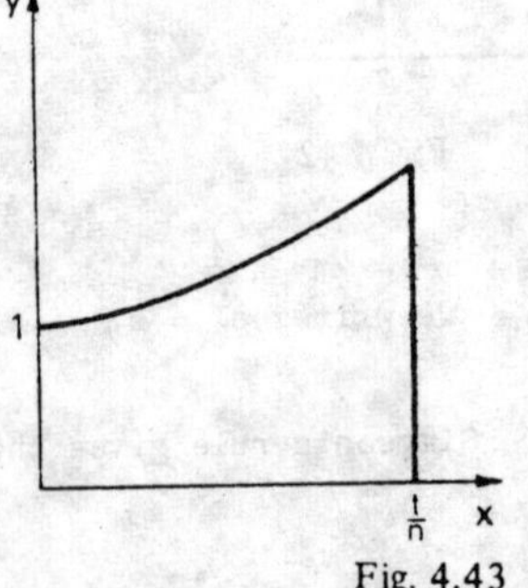

Fig. 4.43

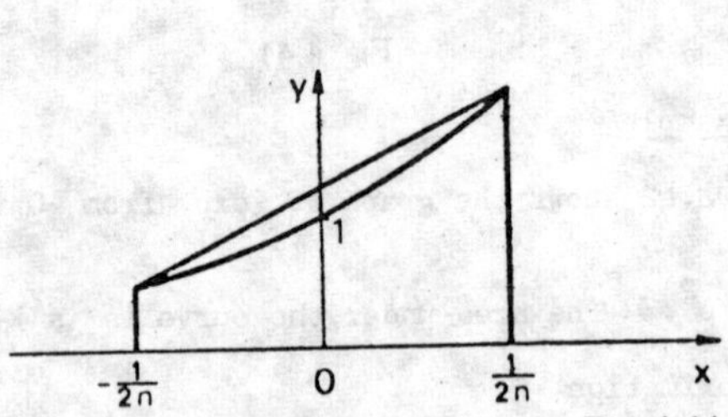

Fig. 4.44

$$\text{(X)} \qquad e^t \doteq \left(\frac{1 + \frac{t}{2n} + \frac{t^2}{12n^2}}{1 - \frac{t}{2n} + \frac{t^2}{12n^2}} \right)^n$$

For $t = 1$ we have the sequence

$$\text{(XI)} \qquad e_n = \left(\frac{1 + \frac{1}{2n} + \frac{1}{12n^2}}{1 - \frac{1}{2n} + \frac{1}{12n^2}} \right)^n$$

which converges very rapidly towards e.

Exercises

7. Print a table of e_n for $n = 1, 2, 2^2, \ldots 2^7$.

8. Calculate $\sqrt{e}$ using (X) and compare with the correct value.

9. Using (X), produce a sequence c_n which converges rapidly to e^{-1}. Estimate e^{-1} by calculating c_n for $n = 1000$.

10. Fig. 4.44 shows the graph of exp from $-\frac{1}{2n}$ to $\frac{1}{2n}$

(a) What is the area under the curve?

(b) What approximation for this is given by the trapezium rule?

(c) Show by equating the two values that

$$a_n = f(n) = \left(\frac{1 + \frac{1}{2n}}{1 - \frac{1}{2n}} \right)^n \qquad \text{is a sequence converging to e.}$$

(d) Show that $f(n) = f(-n)$.

Remark We have already met the sequence $f(n)$ in 3.6. There we found its limiting value with the Romberg rule.

b) Sine and cosine

A particle with mass 1 moves along the y axis (Fig. 4.45). The state of the particle is determined by its position $y(t)$ and its speed $y'(t) = x(t)$. Let the particle start at the origin at time $t = 0$ with speed 1, i.e. $y(0) = 0$, $y'(0) = x(0) = 1$. If a force $-y(t)$ is exerted on the particle then we have $y''(t) = x'(t) = -y(t)$ and its motion is called simple harmonic oscillation. It is described by the following system of differential equations:

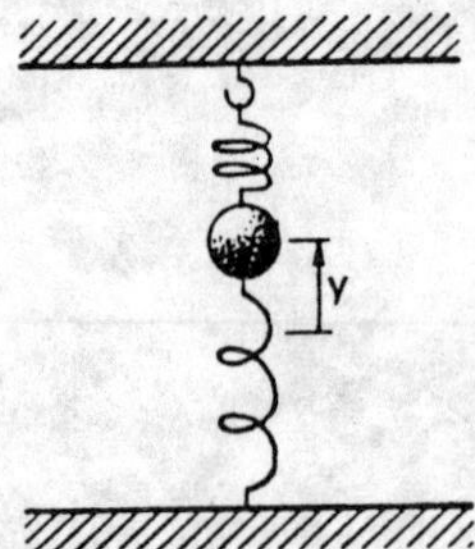

Fig. 4.45

(A)
$$\begin{aligned} x'(t) &= -y(t), & x(0) &= 1 \\ y'(t) &= x(t), & y(0) &= 0 \end{aligned}$$

By integrating from 0 to t we obtain the equivalent system of integral equations

(B)
$$\begin{aligned} x(t) &= 1 - \int_0^t y(u)\,du \\ y(t) &= \int_0^t x(u)\,du \end{aligned}$$

We apply the following integration program to (B)

(C)
$$\begin{aligned} & x(t) \leftarrow 1 \\ \rightarrow\ & y(t) \leftarrow \int_0^t x(u)\,du \\ & x(t) \leftarrow 1 - \int_0^t y(u)\,du \\ & \text{prt } x(t),\ y(t) \end{aligned}$$

It produces the two well-known series for cosine and sine

(D)
$$\begin{aligned} x(t) &= 1 - \frac{t^2}{2!} + \frac{t^4}{4!} - \frac{t^6}{6!} + \dots \\ y(t) &= t - \frac{t^3}{3!} + \frac{t^5}{5!} - \frac{t^7}{7!} + \dots \end{aligned}$$

Exercises

11. Write a program which when x is input will output $\sin x$. Since $\sin(x + 2\pi) = \sin x$, input values should be reduced modulo 2π by using $x \leftarrow x - 2\pi\left[\frac{x}{2\pi}\right]$. The ith term t of the sine series (D) is obtained from the preceding term by the substitution $t \longleftarrow -\frac{tx^2}{2i(2i+1)}$. For each new term a test is made whether $|t| < 10^{-12}$. If the answer is 'yes', the calculation stops. Otherwise the new term is added to the sum, and i increased by 1 ($i \leftarrow i + 1$). The program will also calculate $\cos x$, since $\sin\left(\frac{\pi}{2} + x\right) = \cos x$. In order to obtain $\cos x$, the input value should be $x + \frac{\pi}{2}$.

12. It is still more advantageous, in the previous exercise, to reduce the input x, mod 2π, so that it lies in the interval $-\pi < x \leq \pi$. Verify that the assignment $x \leftarrow x + 2\pi\left[\frac{\pi - x}{2\pi}\right]$ performs this.

4.42 Numerical integral of differential equations

As a preparation for the general case we consider once more the two examples in 4.41.

1. Example

We seek a function $y(x)$ with the property

(1) $$y'(x) = y(x)$$

(2) $$y(x_0) = y_0 .$$

Starting at the known point (x_0, y_0) we go a small step h to the point x_1 (Fig. 4.46), and attempt to calculate an approximate value of $y_1 = y(x_1)$. To do this we integrate (1) from x_0 to x_1. The left side gives $y_1 - y_0$. i.e.

(3) $$y_1 - y_0 = \int_{x_0}^{x_1} y(x)\, dx.$$

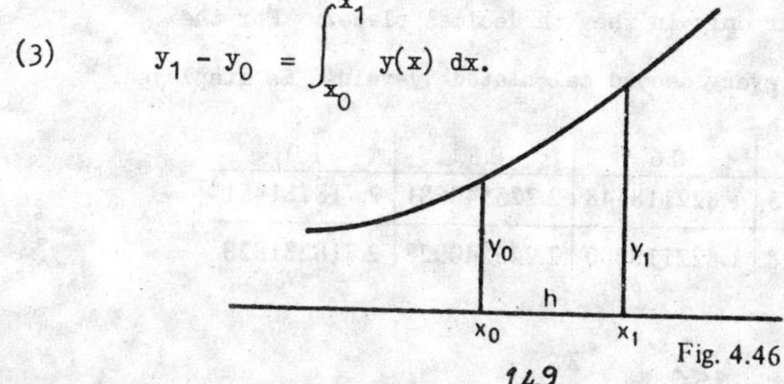

Fig. 4.46

a) We find an estimate for the right hand side using the trapezium rule.

(4) $$y_1 = y_0 + \frac{h}{2}(y_0 + y_1)$$

or

(5) $$y_1 = \frac{2+h}{2-h} y_0 .$$

Both the last two equations are only approximately true. The approximation is the better the smaller h is. The same relationship (5) holds between y_n and y_{n+1}, where $y_n = y(x_n)$ and $x_n = x_0 + n_h$. i.e.

(6) $$y_{n+1} = \frac{2+h}{2-h} y_n .$$

In order to demonstrate the accuracy of the approximation we start with $x_0 = 0$, $y_0 = 1$, $h = 0{\cdot}1$ and calculate table 4.47 using (6).

x	0.0	0.1	0.2	0.3	0.4	0.5	0.6	0.7	0.8	0.9	1.0
y	1.0000	1.1053	1.2216	1.3502	1.4923	1.6494	1.8230	2.0149	2.2270	2.4615	2.7206
e^x	1.0000	1.1052	1.2214	1.3499	1.4918	1.6487	1.8221	2.0138	2.2255	2.4596	2.7183

Fig. 4.47

The third row gives the exact solution $y = e^x$ for comparison.

(b) Considerably better approximations are given by the Hermite rule. Since $y'_0 = y_0$, $y'_1 = y_1$,

$$y_1 - y_0 = \int_{x_0}^{x_1} y(x)\,dx = \frac{h}{2}(y_0 + y_1) + \frac{h^2}{12}(y_0 - y_1)$$

Solving for y_1 gives

$$y_1 = \frac{12 + 6h + h^2}{12 - 6h + h^2} y_0 \quad \text{and so} \quad y_{n+1} = \frac{12 + 6h + h^2}{12 - 6h + h^2} y_n.$$

If we again put $x_0 = 0$, $y_0 = 1$, $h = 0{\cdot}1$ we obtain table 4.48. Departures from the exact solution appear only in the 7th decimal place. For the purpose of this summary only every second calculated y-value is displayed.

x	0	0.2	0.4	0.6	0.8	1.0
y	1	1.221402724	1.491824615	1.822118648	2.225540681	2.718281451
e^x	1	1.221402758	1.491824698	1.822118800	2.225540928	2.718281828

Fig. 4.48

2. Example

The functions sin x and cos x can be defined by

$$s' = c, \quad c' = -s, \quad s(0) = 0, \; c(0) = 1$$

The trapezium rule gives the following relationship between values one step h apart:

$$s_{n+1} = s_n + \int_{x_n}^{x_{n+1}} c(x)dx = s_n + \frac{h}{2}(c_n + c_{n+1})$$

$$c_{n+1} = c_n - \int_{x_n}^{x_{n+1}} s(x)dx = c_n - \frac{h}{2}(s_n + s_{n+1})$$

By solving for s_{n+1} and c_{n+1}, we obtain the recursion equations

(7)
$$s_0 = 0 \qquad c_0 = 1$$
$$s_{n+1} = \frac{(4 - h^2)s_n + 4hc_n}{4 + h^2}, \quad c_{n+1} = \frac{(4 - h^2)c_n - 4hs_n}{4 + h^2}$$

We write $h = \frac{\pi}{6n}$, and calculate the values s_n and c_n using (7) for n = 10, 50, 100, 500. . . In table 4.49 the exact values $\sin\frac{\pi}{6} = 0{\cdot}5$ and $\cos\frac{\pi}{6} = \frac{\sqrt{3}}{2} = 0.86602\,54038$ are given.

n	s_n	c_n
10	0.499896443	0.866085185
50	0.499995856	0.866027796
100	0.499998964	0.866026002
500	0.499999959	0.866025428
1000	0.499999988	0.866025408

$\sin\frac{\pi}{6} = 0.500000000, \quad \cos\frac{\pi}{6} = 0.866025404$

Fig. 4.49

Exercises

1, Write programs to print the tables 4.47 to 4.49.

2. a) Verify that in table 4.49 the extrapolated values

$$s'_{100} = s_{100} + \frac{s_{100} - s_{50}}{3} \quad \text{and} \quad c'_{100} = c_{100} + \frac{c_{100} - c_{50}}{3}$$

give the limiting values $\frac{1}{2}$ and $\frac{\sqrt{3}}{2}$ correct to 10 decimal places.

b) Calculate s_{20} and c_{20} and examine how accurate are the approximations

$$s'_{20} = s_{20} + \frac{s_{20} - s_{10}}{3}, \quad c'_{20} = c_{20} + \frac{c_{20} - c_{10}}{3}$$

to the respective limits $\frac{1}{2}$ and $\frac{\sqrt{3}}{2}$.

After this preparation we consider the following <u>Initial value problem</u>

> Determine the function y which satisfies
>
> (8) $y' = f(x, y)$
>
> and the <u>initial condition</u>
>
> (9) $y(x_0) = y_0$

We start at (x_0, y_0) and move about in the plane subject to the conditions (8) which specify exactly the direction of motion at any point.

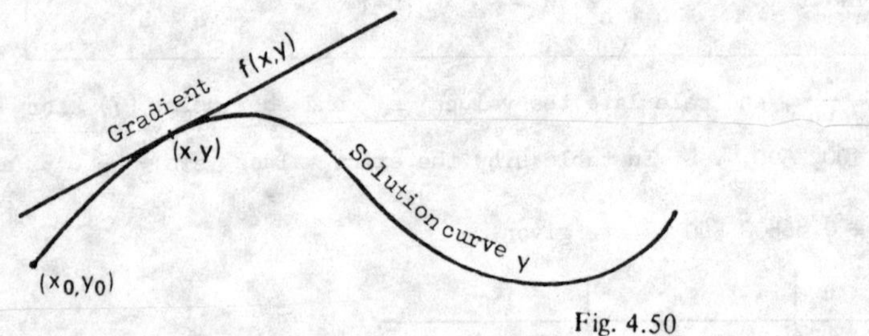

Fig. 4.50

But $y' = f(x, y)$ demands a steady alteration of course which we cannot keep to exactly. So we have to be satisfied with one of the following approximation methods.

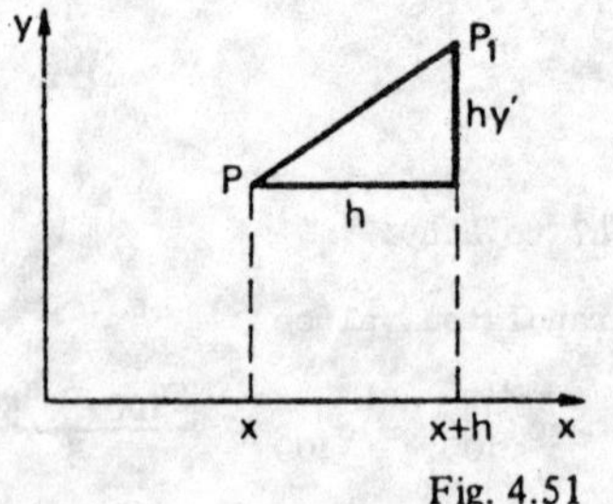

Fig. 4.51

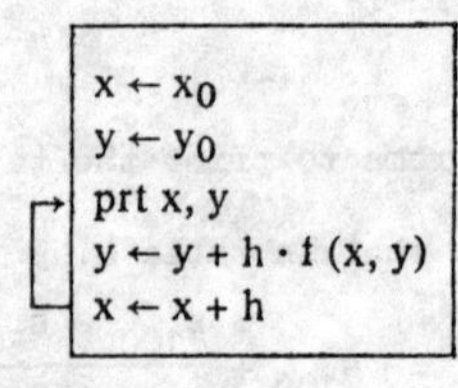

Fig. 4.52

1. <u>The method of the Euler-polygon</u>

Suppose that we are now at P(x, y). We go a small step along the straight line with gradient $y' = f(x, y)$ to the next point $P_1 = (x + h, y + hy')$ (Fig. 4.51). Then P_1 replaces P and the step is repeated. In this way we approximate the solution curve with a 'polygonal arc'. Fig. 4.52 shows the program. We shall see in the exercises that the error is proportional to h.

2. <u>The trapezium method</u>

We want to improve the crude Euler method. Suppose we are at (x_n, y_n). We wish to determine the next point (x_{n+1}, y_{n+1}), where $x_{n+1} = x_n + h$. Integration of $y' = f(x, y)$ from x_n to x_{n+1} gives

$$(10) \qquad y_{n+1} = y_n + \int_{x_n}^{x_{n+1}} f(x,y)\,dx.$$

The integral is approximated using the trapezium rule

$$(11) \qquad y_{n+1} = y_n + \frac{h}{2}\left(f(x_n,y_n) + f(x_{n+1}, y_{n+1})\right)$$

In the examples 1 and 2 we solved these equations for y_{n+1}. This is easy when f(x,y) is linear in y. Usually a solution is not possible. So we estimate y_{n+1} crudely with

$$(12) \qquad \hat{y}_{n+1} = y_n + hy'_n = y_n + hf(x_n, y_n)$$

This value is called the <u>predictor</u>, because it predicts the value of y_{n+1}. If $\hat{y}_{n+1}$ is substituted for y_{n+1} in the right hand side of (11), we obtain the improved approximation.

$$(13) \qquad y_{n+1} = y_n + \frac{h}{2}(f(x_n, y_n) + f(x_{n+1}, \hat{y}_{n+1})).$$

This is the <u>corrector</u>, since it corrects the approximation $\hat{y}_{n+1}$. Fig. 4.53 shows the program.

3. <u>The mid-point method</u>

The integral in (10) can be approximated by the mid-point rule. We have

$$y_{n+1} = y_n + hf(x_n + \tfrac{h}{2}, y_m) \tag{14}$$

where $y_m = y(x_n + \frac{h}{2})$ is unknown and may be crudely estimated by

$$y_m = y_n + \frac{h}{2} y'_n = y_n + \frac{h}{2} f(x_n, y_n).$$

From this we obtain the approximation

$$y_{n+1} = y_n + hf\left(x_n + \frac{h}{2}, y_n + \frac{h}{2} f(x_n, y_n)\right). \tag{15}$$

Fig. 4.54 shows the program.

```
   x ← x0
   y ← y0
┌→ prt x, y
│  a ← x + h
│  b ← y + hf (x, y)
│  y ← y + h/2 (f (x, y) + f (a, b))
└─ x ← a
```

Fig. 4.53

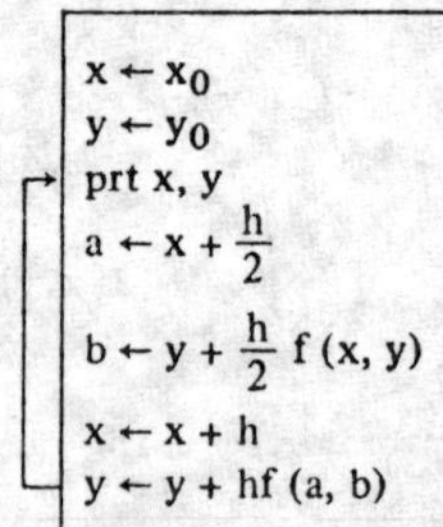

Fig. 4.54

<u>Exercise</u>:

3. We apply the three methods to $y' = \frac{x}{y}$, $y(0) = 1$.

   a) Let $h = 0{\cdot}1$. Print the points (x,y) from $x = 0$ to $x = 1$ using each of the programs in Fig. 4.52 - 4.54 and compare with the exact solution $y = \sqrt{1 + x^2}$.

   b) Solve the exercise a) for the step-length $h = 0{\cdot}01$. The points (x,y) should be printed only when $10x = [10x]$ i.e. for $x = 0, 0{\cdot}1, 0.2, \ldots \ldots 1{\cdot}0$.

   c) If we write $h \leftarrow \frac{h}{10}$ the error in the Euler method is reduced about 10 fold, and in the two other methods about 100 fold. Verify this with the aid of the data from a) and b).

4.4.3. Simulation of dynamic processes

The computer is a simulator of processes, of deterministic and stochastic processes. This is the most important use of computers. We discuss four examples of deterministic dynamic processes.

1st Example. A problem of pursuit

Pursuit problems were first studied by Leonardo da Vinci. They are ideal computer problems. Computationally they are hopeless. But they can be described by very simple programs. They need only the theorem of Pythagoras and the ideas of similarity, or still better, a little vector geometry. They can also be solved graphically, in which case no prior knowledge is needed. The simplest version runs as follows:

(a) A hare runs Northwards. A fox pursues him by running always directly towards the hare. Draw the curve of pursuit.

Suppose the hare is initially at $O(0,0)$ and the fox is at $F_o(x_o, y_o)$.

The fox uses the following algorithm:

1. Locate the current position of the hare.
2. Run in that direction for a distance q.

   (The hare runs a distance p in the same time period).
3. Go to step 1.

The fox cannot alter his direction of motion continuously, since he moves in jumps. The smallest possible q for him is the length of one jump. The pursuit-process is therefore discrete and can easily be simulated by the computer. Among the most interesting is the case $p = q$. This special case is shown in Fig. 4.55. The construction can be understood without explanation. The numbers give the times at which the fox and hare are to be found at the marked points.

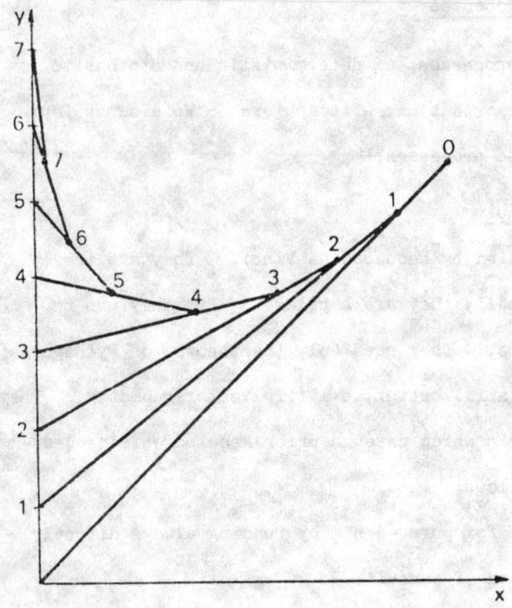

Fig. 4.55

Let $\vec{F} = \binom{x}{y}$ and $\vec{H} = \binom{0}{z}$ be the current position-vectors of fox and hare. Then $\overrightarrow{FH}$ is $\binom{-x}{z-y}$ and $d = |\overrightarrow{FH}| = \sqrt{x^2 + (y-z)^2}$.

The new position after one jump is obtained by the substitutions

$$\vec{F} \leftarrow \vec{F} + \frac{q}{d}\overrightarrow{FH}, \quad \vec{H} \leftarrow \vec{H} + p\binom{0}{1}. \qquad \text{(See Fig. 4.56)}$$

Fig. 4.57 shows the program in coordinates. If p and q are chosen sufficiently small the curves of pursuit are almost smooth.

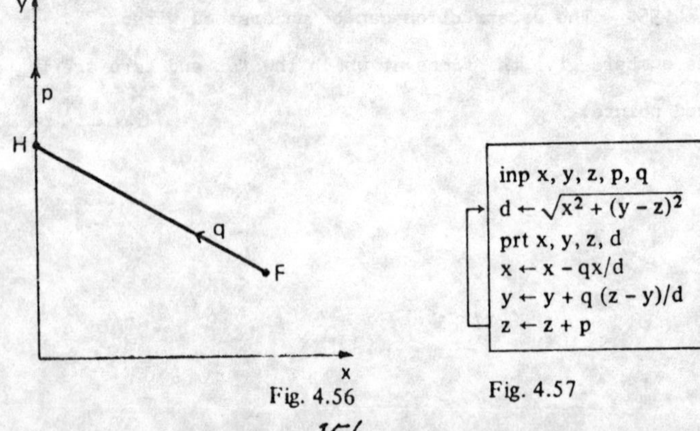

Fig. 4.56 Fig. 4.57

(b) Orthogonal flight. The fox always runs in the direction of the hare. The hare tries to dodge by always running at rightangles to the direction FH. The position vectors of F and H are now

$$\vec{F} = \binom{x}{y} \quad \text{and} \quad \vec{H} = \binom{u}{v}$$

and after one jump $\vec{F}_1 = \binom{x_1}{y_1}$ and $\vec{H}_1 = \binom{u_1}{v_1}$.

If we write $d = |\vec{FH}|$ and note that by a rotation about the origin of 90°, the vector $A = \binom{a}{b}$ becomes $\binom{-b}{a}$, we have, from Fig. 4.58

$$\vec{FH} = \binom{u-x}{v-y}, \quad \vec{FF}_1 = \frac{q}{d}\binom{u-x}{v-y}, \quad \vec{HH}_1 = \frac{p}{d}\binom{-v+y}{u-x}$$

$$\vec{F}_1 = \vec{F} + \frac{q}{d}\binom{u-x}{v-y}, \quad \vec{H}_1 = \vec{H} + \frac{p}{d}\binom{-v+y}{u-x}$$

Fig. 4.59 shows the program.

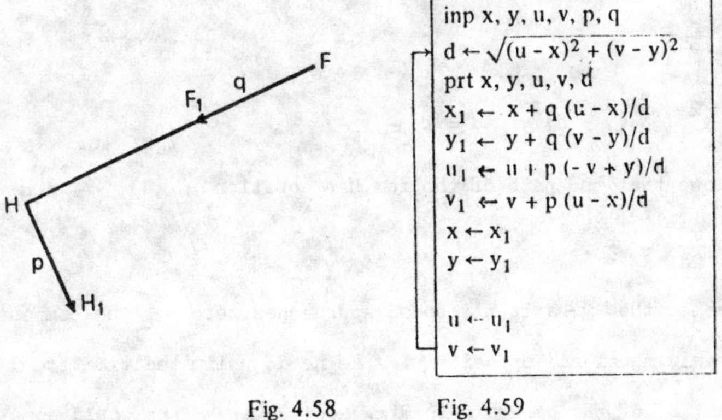

Fig. 4.58 Fig. 4.59

(c) The problem a) can be radically simplified. We represent the path of the fox in a moving coordinate system, in which the hare stays at 0. The fox has two velocities: in the direction of 0, and in the negative y - direction of constant magnitude q. If q = p, the resultant velocity bisects the angle between F_o0 and F_oL. It follows that in the case p = q the fox runs in a parabola with focus 0 and directrix 1.

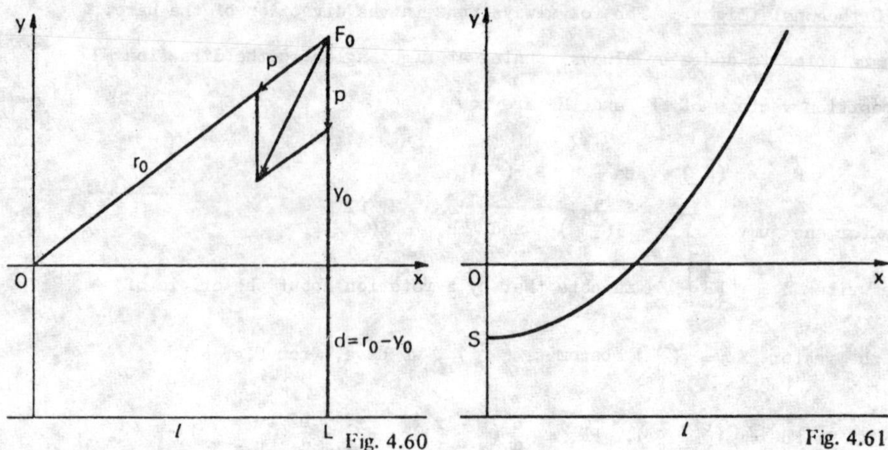

Fig. 4.60 Fig. 4.61

The equation of the path is easily calculated without prior knowledge. Initially the fox is the same distance from 0 and 1 namely $r_o = \sqrt{x_o^2 + y_o^2}$. As he approaches both 0 and 1 with the same speed, the relation $\overline{FO} = \overline{FL}$ remains true and so

$$x^2 + y^2 = (y + d)^2, \quad d = r_o - y_o$$

From that it follows that the path of the fox has equation

$$y = \frac{x^2}{2d} - \frac{d}{2}.$$

As the fox approaches the y-axis his speed approaches zero [the component velocities are nearly equal and opposite]. He needs unlimited time to reach the vertex S of the parabola in Fig. 4.61. At S he would come to rest i.e, the distance between fox and hare tends to

$$\overline{OS} = \frac{d}{2} = \frac{r_o - y_o}{2} = \frac{\sqrt{x_o^2 + y_o^2} - y_o}{2},$$

where (x_o, y_o) is the starting position of the fox. Fig. 4.62 gives a program for the path of the fox in a moving coordinate system. It is written for the general case $p \neq q$.

```
inp x, y, p, q
d ← √(x² + y²)
prt x, y, d
x ← x - qx/d
y ← y - qy/d - p
```

Fig. 4.62

Exercise

1. a) Complete the program in Fig. 4.57 and translate it into BASIC. Use initial conditions $x = 40$, $y = 30$, $z = 0$, $p = q = 1$. The position of the fox and the hare should be printed only after 10, 20, . . . 160 jumps.

   b) Alter the program so that only the distance d is printed, after 100, 200, . . . 1000 jumps. Compare with the limiting distance $d_\infty = 10$.

   c) Choose $p = q = 2$ and print d after 50, 100, . . . 500 jumps. Compare with b).

2. a) Draw as in Fig. 4.55, the paths of fox and hare for the case of orthogonal flight with $u = v = 0$, $x = 10$, $y = 0$, $p = q = 1$.

   b) Write a BASIC program for Fig. 4.59 which executes the PRT-statement only after 100, 200, . . . 2000 jumps. Input $u = v = 0$, $x = 10$, $y = 0$, $p = q = 0{\cdot}01$ and compare with the crude approximation in a).

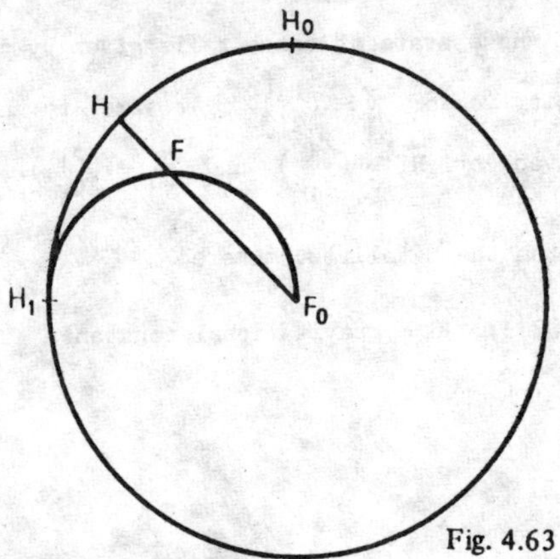

Fig. 4.63

3. Fig. 4.63 shows a circular hedge with radius 1. A hare starts at H_o and runs along the hedge anticlockwise. A fox starts at F_o and runs along always directly towards the hare.

   a) Construct and calculate the path of the fox for $p = q$.

   [b)] The fox alters his strategy. He now runs so that he always stays on the line F_oH. Show that he catches the hare at H_1 where $\measuredangle H_oF_oH_1 = 90^\circ$.

2nd Example. Planetary motion.

Analysis is a tool for the investigation of processes. The state of a process can be represented by a vector $\vec{X}(t)$. The set of all possible states is called the process-space or phase-space. One may regard $\vec{X}(t)$ as the position vector of a particle which moves in the phase-space (Fig.4.64). From physical considerations one can indicate the velocity $\dot{\vec{X}}(t) = \vec{v}(t)$ of the particle. This is the local law of the evolution of the process. From this local law we can reconstruct the history of the process and predict its future development. This is brought about by integration of the local law i.e. a differential equation. With the aid of computers it is possible to simulate directly the local law of a process. This we can manage completely without knowledge of analysis. We shall show this for planetary motion as an example.

Let h be a short interval of time. We consider a heavy body of mass M at O, and a light satellite of mass m, whose state at time t is given by the position vector $\vec{R} = \binom{x}{y}$ and velocity vector $\vec{v} = \binom{p}{q}$. At time $t + h$ let the state of the satellite be given by $\vec{R}_1 = \binom{x_1}{y_1}$ and $\vec{v}_1 = \binom{p_1}{q_1}$.

A gravitational force $\vec{F}$ is exerted on the satellite, where

$$\vec{F} = -\frac{GmM}{r^3}\vec{R}, \quad r = |\vec{R}|, \quad G = \text{gravitational constant.}$$

This gives it an acceleration

$$\vec{b} = \frac{\vec{v}_1 - \vec{v}}{h}$$

The dynamical law $\vec{F} = m\vec{b}$ [Newton] gives, after a short calculation,

$$\vec{v}_1 = \vec{v} - \frac{ah\vec{R}}{r^3}, \quad \text{where} \quad a = GM.$$

Also we have

$$\vec{R}_1 = \vec{R} + \vec{v}\,h$$

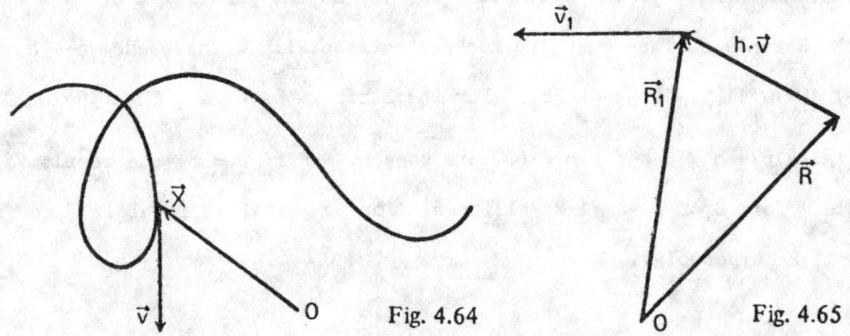

Fig. 4.64

Fig. 4.65

The kernel of the program, which simulates the motion of the satellite, consists of the following four lines:

$\vec{R}_1 \leftarrow \vec{R} + h\vec{v}$	(calculate new position)
$\vec{v}_1 \leftarrow \vec{v} - \frac{ah\vec{R}}{r^3}$	(calculate new velocity)
$\vec{R} \leftarrow \vec{R}_1$	(Update position and velocity. The old state is replaced by
$\vec{v} \leftarrow \vec{v}_1$	the new.)

Fig. 4.66 shows the program. It follows the satellite for t time intervals. The position and velocity of the satellite as well as its distance from 0 are printed for each time interval.

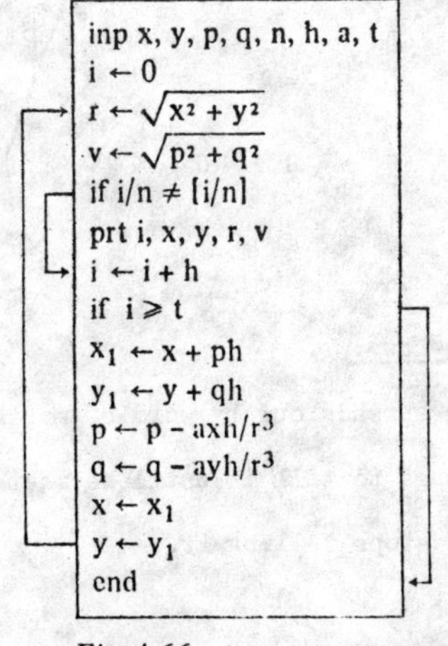

Fig. 4.66

We take the centre of the Earth as origin. At the point $x = 0$, $y = 6371$ km. on the Earth's surface we fire rockets horizontally with speeds $p = 8$, 10, 10·5, 10·7, 11·2, 12·5 km/s. For the Earth we have $a = GM = 398200$ km^3/s^2.

In Fig. 4.67 $h = 5$ sec, $n = 600$ are chosen. For $p = 8$ the result is a circle, for $8 < p < 11{\cdot}2$ ellipses, for $p = 11{\cdot}2$ a parabola and for $p > 11{\cdot}2$ hyperbolas.

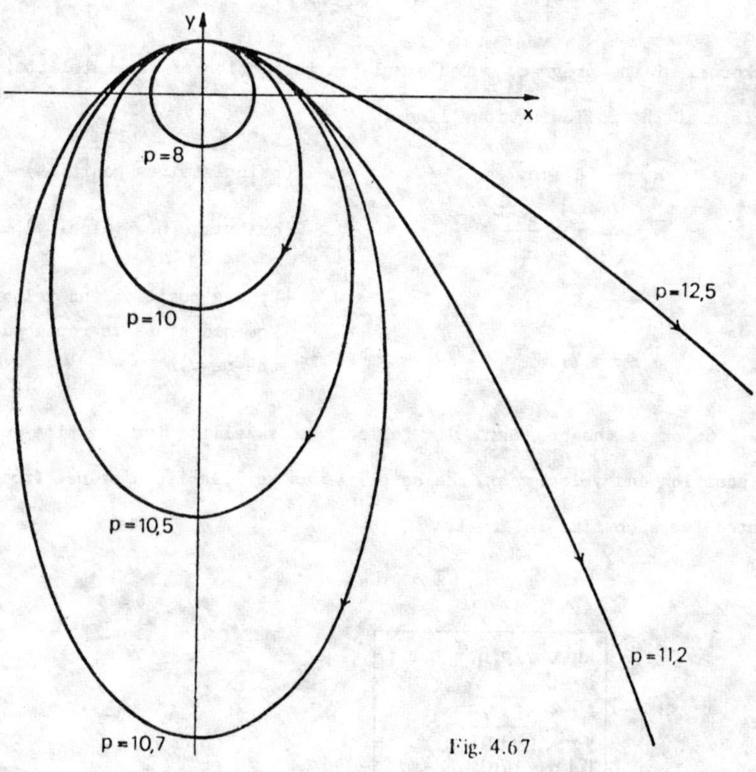

Fig. 4.67

<u>3rd Example. A rumour spreads through the town.</u>

In a town with $n + 1$ inhabitants a rumour spreads out by word of mouth. Each person who has heard the rumour continues to tell it until he meets someone who has already heard it. Then he stops telling it.

We divide the inhabitants into three groups X, Y, Z with x, y, z members respectively.

X contains all those who have not yet heard the rumour ('susceptible').

Y contains all those who are actively spreading the rumour ('infectious').

Z contains all who no longer spread the rumour ('immune').

Initially we have $x = n$, $y = 1$, $z = 0$, and it is always true that $x + y + z = n + 1$.

We must consider three types of transition:

$$\text{"Infection"} \quad (x,y,z) \leftarrow (x - 1,\ y + 1,\ z)$$

$$\text{"Immunisation"} \begin{cases} (x,y,z) \leftarrow (x,\ y - 1,\ z + 1) \\ (x,y,z) \leftarrow (x,\ y - z,\ z + 2) \end{cases}$$

After a short time interval h, x and y should be replaced by x_1 and y_1 respectively. Infections occur through XY-encounters. The number of possible XY pairs is xy. The change in x is proportional to xy. That is

$$x_1 = x - pxy.$$

Immunisations occur through YZ- and YY-encounters. There are yz pairs of the first kind and $\frac{y(y - 1)}{2}$ pairs of the second kind. Since YY-encounters increase z by 2, the number of immunisations is proportional to $yz + \frac{2y(y - 1)}{2}$. Since $x + y + z = n + 1$, this is $y(n - x)$. That is, in time interval h, y is increased by pxy (inflow from X) and reduced by $py(n - x)$ (outflow to Z). Therefore

$$y_1 = y - py(n - 2x).$$

From $x = n$ to $x = \frac{n}{2}$, y increases; below $x = \frac{n}{2}$, y decreases. As soon as $y < 1$ becomes true the rumour will stop. We are interested in the corresponding x value. This gives the number of people who never hear the rumour. The program in Fig. 4.68 prints the proportion $\frac{x}{n}$ of people who never hear the rumours.

```
10 INPUT N, P
20 X = N
30 Y = 1
40 X1 = X - P * X * Y
50 Y = Y - P * Y * (N - 2 * X)
60 X = X1
70 IF Y >= 1 THEN 40
80 PRINT X/N
90 END
```

Fig. 4.68

Exercise

4. Run the program for

   a) $n = 100, \quad p = 10^{-4}$

   b) $n = 1000, \quad p = 10^{-4}$

   c) $n = 100, \quad p = 10^{-3}$

   d) $n = 10, \quad p = 10^{-3}$

   The results are surprising.

4th Example. Eat and be eaten. The struggle for existence

On an island there is an arbitrarily large amount of grass. There are also hares, who eat grass, and foxes who eat hares. At time t let there be $x(t)$ tons of hares and $y(t)$ tons of foxes. The state of the island is determined by the point $\vec{P}(t) = (x(t), y(t))$ in the phase-space. From biological considerations we have for $\vec{P}'(t)$

$$(1) \quad \begin{cases} \dfrac{dx}{dt} = ax - bxy \\[2ex] \dfrac{dy}{dt} = -cy + dxy \end{cases}$$

Reason : If there were no foxes $(y = 0)$, the hares would increase at a rate $\frac{dx}{dt} = ax$. If there were only foxes $(x = 0)$, they would die off at a rate $\frac{dy}{dt} = -cy$. If both foxes and hares are present, the number of encounters between them is proportional to xy. The number of hares eaten in any time interval is therefore proportional to xy. The hares which are eaten serve as nourishment for the foxes. This explains the terms $- bxy$ and dxy in (1)

We draw the path of P for different starting values using the Euler-method. We also choose $a = b = 2$ and $c = d = 1$. In Fig. 4.69, h is a small time interval. The program traces the path of P from time 0 to time t. Time and position are printed-out at time 0, n, 2n, 3n, . . .

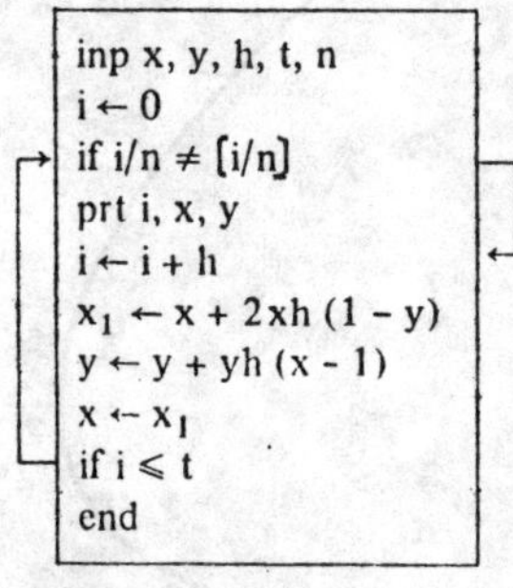

```
inp x, y, h, t, n
i ← 0
if i/n ≠ [i/n]
prt i, x, y
i ← i + h
x1 ← x + 2xh (1 - y)
y ← y + yh (x - 1)
x ← x1
if i ≤ t
end
```

Fig. 4.69

We put $h = 0{\cdot}02$, $t = 6$, $n = 0{\cdot}2$ and start at time 0 at the points

I $x = y = 0{\cdot}25$ II $x = y = 0{\cdot}5$ III $x = y = 0{\cdot}75$.

Fig. 4.70 shows the curves which arise. We see a periodic variation in the populations, neither of which dies out. The periodic time is almost independent of the starting point. For the curves I, II, III it amounts to about 5·7, 4·8 and 4·5 respectively. The position of the point P on the curves is shown for times 0, 1, 2, 3, 4, 5 (and for 6 also on curve I). The curves are not quite closed, since the Euler method gives too coarse an approximation. In II one can see that the point 5 does not lie on the curve.

The point G(1, 1) is the equilibrium point. If we start at G, the values x and y remain constant. It can be shown that in the neighbourhood of G the curves are approximate ellipses which are described by P in the time $\pi\sqrt{2} \approx 4{\cdot}443$. (see [6]).

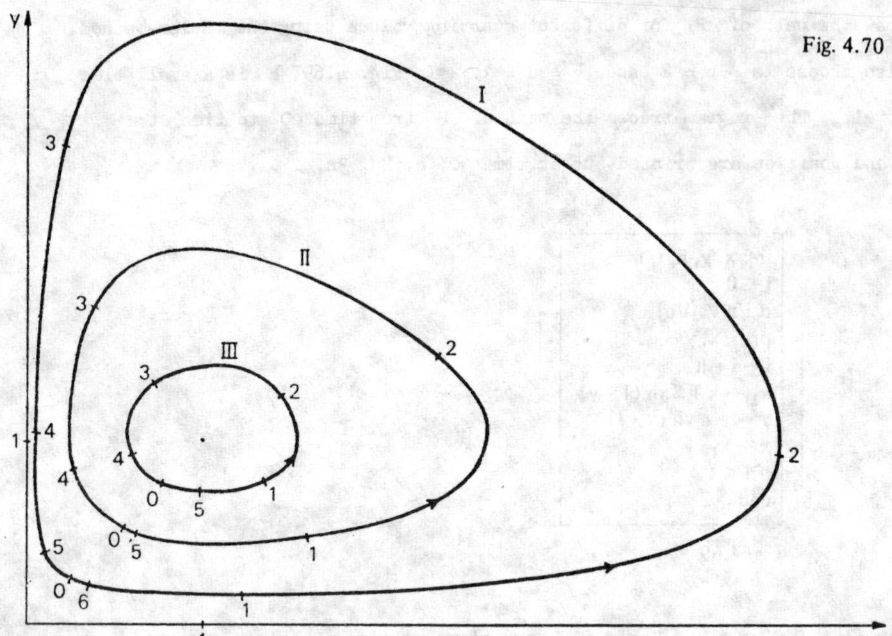

Exercises

5. In (1) Let $a = b = 2$, $c = d = 1$. From $y' = \dfrac{\frac{dy}{dt}}{\frac{dx}{dt}}$. it follows that

(2) $y' = f(x, y) = \dfrac{2x(1 - y)}{y(x - 1)}$.

Determine the curves I to III by integration of (2) using

a) the trapezium rule (Fig. 4.53) b) the mid-point rule (Fig. 4.54).

6. A simple circle algorithm (for readers with plotters) [HRG †]

Using the relations

$$x_{n+1} = x_n - ay_n$$

$$y_{n+1} = y_n + ax_{n+1} \qquad |a| \leqslant 2$$

one obtains sequences of points which form smooth curves.

For $|a| < 0{\cdot}02$ they are nearly circles; actually they are almost-circular ellipses which are symmetrical about $y = x$ for $a > 0$ (and about $y = -x$ for $a < 0$). As a increases in magnitude they become polygons which are inscribed in ellipses. Draw the curves for

(a) $x = y = 30$, $a = 0{\cdot}02, 0{\cdot}1, 0{\cdot}2, 0.5, 1, 1{\cdot}5, 1{\cdot}6, 2.$

(b) $x = y = 20$, $a = -0{\cdot}02, -0{\cdot}1, -0{\cdot}5, -1, -1{\cdot}5, -1{\cdot}6, 2.$

(c) Choose $a = 2 \sin \frac{360^\circ}{k}$ and $k = 4, 5, 6, 7, 8, 9, 10, 11, 12\ldots$

What happens?

7. Solar wind. (For readers with plotters [HRG]).

The U.S. Echo-satellites are very light giant balloons of 30-40 m diameter. Their motion is strongly influenced by the "solar-wind"; this gives them a constant acceleration s which we will take to be in the positive x-direction. Line 11 in Fig. 4.66 should now read $p \leftarrow p + (s - \frac{ax}{r^3})\, h$. Experiment with different values of s and observe the fantastic orbits.

4.5 The haromonic series

The series

$$H_n = 1 + \frac{1}{2} + \frac{1}{3} + \ldots + \frac{1}{n}, \quad n = 1, 2, 3, \ldots$$

is called the harmonic series. It is important because it appears in numerous applications. It is easy to show that H_n diverges. Yet the sequence grows unusually slowly, as we shall shortly see. Calculation of the members of the sequence even by computer can thus be very time consuming. In order to calculate $H_{1000000}$ a pocket calculator must be left running all night. We set ourselves the goal of discovering a good approximation formula by using numerical experimentation with the computer or pocket-calculator.

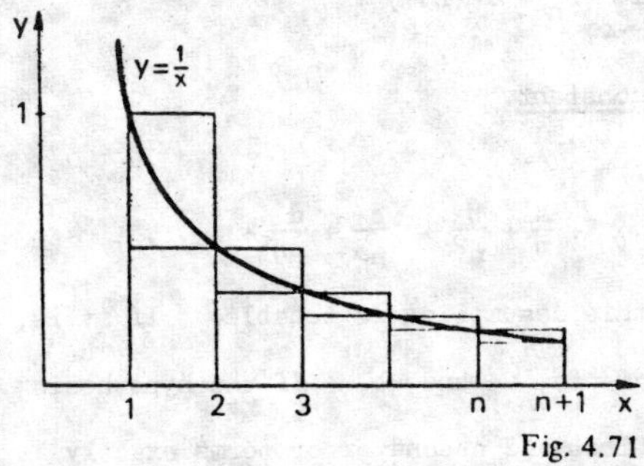

Fig. 4.71

Using Fig. 4.71,

the area of the upper 'staircase' from 1 to n+1 $= H_n$,

the area of the lower 'staircase' from 1 to n $= H_n - 1$,

the area under the curve from 1 to (n + 1) $= \ln(n + 1)$,

and the area under the curve from 1 to n $= \ln(n)$.

Thus

$$H_n > \ln(n + 1) > \ln(n) \quad \text{and} \quad H_n - 1 < \ln(n)$$

or

$$\ln(n) < H_n < \ln(n) + 1.$$

We consider the sequence

$$c_n = H_n - \ln(n)$$

Then $c_n > 0$. We show that c_n decreases monotonely. In fact (Fig. 4.72)

$$c_{n+1} - c_n = H_{n+1} - H_n - (\ln(n + 1) - \ln(n)) = \frac{1}{n+1} - \ln(1 + \frac{1}{n})$$

= area of the rectangle ABCD - area of the 'curved trapezium' ABCE < 0

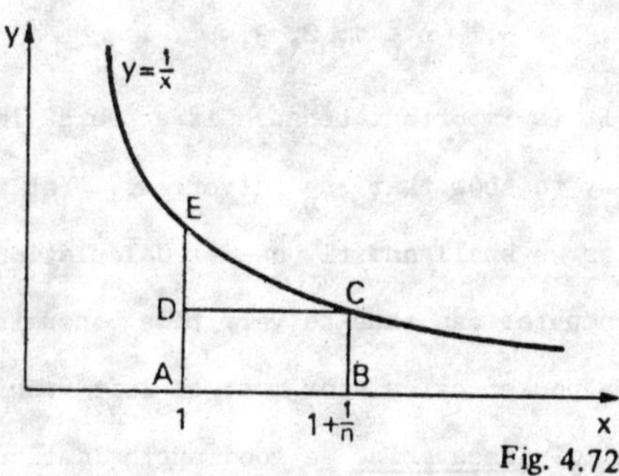

Fig. 4.72

The monotone decreasing sequence is bounded below and so has a limit

$$\gamma = \lim_{n \to \infty} c_n = \lim_{n \to \infty} (H_n - \ln(n)) .$$

The limit γ is called Euler's constant.

We make the initial assumption

(1) $$c_n = H_n - \ln(n) = \gamma + \frac{a}{n} + \frac{b}{n^2} + \frac{c}{n^3} + \frac{d}{n^4} + \ldots$$

and test experimentally whether this assumption is tenable. If it is, we will try to determine the coefficients a and b. If the hypothesis (1) is correct we can eliminate the first and second error terms exactly as on page

144 and so obtain a sequence which converges much more quickly. From (1) it follows that

$$(2) \qquad c'_{2n} = c_{2n} + \frac{1}{2^1 - 1}(c_{2n} - c_n) = \gamma + \frac{b'}{n^2} + \frac{c'}{n^3} + \frac{d'}{n^4} + \dots$$

$$(3) \qquad c''_{2n} = c'_{2n} + \frac{1}{2^2 - 1}(c'_{2n} - c'_n) = \gamma + \frac{c''}{n^3} + \frac{d''}{n^4} + \dots$$

We calculate the sequence $c_1, c_2, c_4, c_8, \dots c_{128}$ from which we easily obtain $c'_2, c'_4, c'_8, \dots c'_{128}$, and $c''_4, c''_8, \dots c''_{128}$ from (2) and (3).

Table 4.73 shows the result

n	c_n	c'_n	c''_n	c'''_n
1	1.0000000000			
2	0.8068528194	0.6137056389		
4	0.6970389722	0.5872251250	0.5783982870	
8	0.6384156012	0.5797922302	0.5773145985	0.5772423526
16	0.6081402710	0.5778649407	0.5772225109	0.5772163718
32	0.5927592926	0.5773783143	0.5772161054	0.5772156784
64	0.5850078203	0.5772563480	0.5772156926	0.5772156651
128	0.5811168286	0.5772258369	0.5772156665	0.5772156648

Fig. 4.73

If (1) proves true, then $\hat{\gamma} = c''_{128} = 0{\cdot}57721\,56665$ is a very good approximation for γ. We can calculate that $c_{64} - \hat{\gamma} \doteq 0{\cdot}0078$, $c_{128} - \hat{\gamma} \doteq 0{\cdot}0039$. Thus by doubling n, $|c_n - \hat{\gamma}|$ is approximately halved. This points to the existence of the error term $\frac{a}{n}$ in (1). Further, $c'_{64} - \hat{\gamma} = 0{\cdot}0000406815$, $c'_{128} - \hat{\gamma} = 0{\cdot}0000101704$. Doubling n therefore almost exactly divides $|c'_n - \hat{\gamma}|$ by 4. This confirms the existence of the error terms $\frac{b}{n^2}$ in (1) and $\frac{b'}{n^2}$ in (2) respectively.

Further, $c''_{16} - \hat{\gamma} = 6{\cdot}844 \times 10^{-6}$, $c''_{32} - \hat{\gamma} = 4{\cdot}389 \times 10^{-7}$, $c''_{64} - \hat{\gamma} = 2{\cdot}61 \times 10^{-8}$

By doubling n, the error $|c''_n - \hat{\gamma}|$ is divided, not by 8 but by 16.

That is, the third order term in (1), (2), (3) is absent.

On the other hand, there does exist a term of order 4. We can eliminate that by

$$c'''_{2n} = c''_{2n} + \frac{1}{2^4 - 1}(c''_{2n} - c''_n)$$

and so obtain the last column of table 4.73. So $c'''_{128} = 0.57721\ 56648$ is an excellent approximation for γ. The exact value is

$$= 0.57721\ 56649\ 01532\ \ldots$$

Now we calculate a and b. From

$$\frac{a}{64} + \frac{b}{64^2} \simeq c_{64} - \gamma \quad = \quad 0.0077921554$$

$$\frac{a}{128} + \frac{b}{128^2} \simeq c_{128} - \gamma \quad = \quad 0.0039011637$$

it follows that $b = -0.083329024 \simeq -0.083333\ldots = -\frac{1}{12}$ and $a = \frac{1}{2}$.

Thus we have

$$H_n = \ln(n) + \gamma + \frac{1}{2n} - \frac{1}{12n^2} + \frac{d}{n^4} + \ldots$$

One can check that d must be very small, somewhat smaller than $\frac{1}{100}$.

Thus, for $n \geqslant 100$, $\left|\frac{d}{n^4}\right| < 10^{-10}$ and we can use the approximation

$$H_n \simeq \ln(n) + \gamma + \frac{1}{2n} - \frac{1}{12n^2}, \text{ where } \gamma = 0.57721\ 56649$$

For $n = 128$ this gives

$$H_{128} \simeq \ln(128) + \gamma + \frac{1}{256} - \frac{1}{12.128^2} = 5.433147093.$$

All digits of this approximation are correct.

Exercises

1. Let $s_n = 1 + \frac{1}{4} + \frac{1}{9} + \ldots + \frac{1}{n^2}$. We wish to calculate $s = \lim_{n\to\infty} s_n$ as precisely as possible. We make the assumption

$$s_n = s + \frac{a}{n} + \frac{b}{n^2} + \frac{c}{n^3} + \frac{d}{n^4} + \ldots$$

By treating this sequence in the same way as the sequence c_n in (1), determine s. Compare with the exact value $\frac{\pi^2}{6}$.

2. Let $s_n = \frac{1}{1.2} + \frac{1}{3.4} + \frac{1}{5.6} + \ldots + \frac{1}{(2n-1)2n}$.

Determine $s = \lim_{s_n \to \infty} s_n$ by calculating a table analogous to that of Fig. 4.73. Compare with the exact value ln2.

4.6 The calculation of e to 250 decimal places

Let $e_n = 1 + \frac{1}{1!} + \frac{1}{2!} + \ldots + \frac{1}{n!}$. From 4.41, Exercise 1, we have $f_n = e - e_n < \frac{n+2}{n+1} \cdot \frac{1}{(n+1)!}$. For $n = 144$, $f_n < 10^{-251 \cdot 9}$.

To calculate e we use Fig. 69 (p 252), where we split the instruction $e \leftarrow 1 + \frac{e}{n}$ into $e \leftarrow \frac{e}{n}$; $e \leftarrow e + 1$. So we have the following program for e_n:

1. Put $e \leftarrow 1$
2. While $n > 0$, put $e \leftarrow \frac{e}{n}$; $e \leftarrow 1 + e$; $n \leftarrow n - 1$.

We calculate in the system with base $B = 10^5$. The single 'digits' will be stored in X(0), X(1), ... X(51). Initially n = 144. Thus

$$\frac{1}{144} = \underbrace{0}_{X(0)} \cdot \underbrace{00694}_{X(1)} \quad \underbrace{44444}_{X(2)} \quad \underbrace{44444}_{X(3)} \ldots \underbrace{44444}_{X(51)} \ldots$$

After that, $X(0) \leftarrow X(0) + 1$ is performed and we have 1·00694 44444 . . . . 44444. This is now divided by 143 exactly as in the base-10 system, except that the number $B = 10^5$ replaces 10. That is, at the ith step, X(I) is divided by 143, the quotient Q becomes X(I) and the remainder R is multiplied by B and added to X(I + 1), and so on, until n = 0. Afterwards we need to dealwith the carrying figures - that is, X(I) must be reduced mod B and $U = \left[\frac{X(I)}{B}\right]$ added to X(I - 1). The last digit X(51) must also be rounded i.e. $U = \left[\frac{X(51)}{B} + 0{\cdot}5\right]$.

The program in Fig. 4.74 carries out this plan.

```
10 DIM X (52)
20 FOR I = 0 TO 52 X (I) = 0
30 B = 1E + 5;  X (0) = 1
40 FOR N = 144 TO 1 STEP - 1
50       FOR I = 0 TO 51
60             Q = INT (X (I)/N);  R = X (I) - Q * N
70             X (I) = Q;  X (I + 1) = X (I + 1) + B * R
80       NEXT I
90       X (0) = X (0) + 1
100 NEXT N
110 X (50) = X (50) + INT (X (51)/B + 0.5)
120 FOR I = 50 TO 1 STEP - 1
130       U = INT (X (I)/B);  X (I) = X (I) - B * U;  X (I - 1) = X (I - 1) + U
140 NEXT I
150 FOR I = 0 TO 50 PRINT X (I);
160 END
```

```
2 71828   18284   59045   23536    2874   71352   66249   77572   47093   69995
  95749   66967   62772   40766   30353   54759   45713   82178   52516   64274
  27466   39193   20030   59921   81741   35966   29043   57290    3342   95260
  59563    7381   32328   62794   34907   63233   82988    7531   95251    1901
  15738   34187   93070   21540   89149   93488   41675    9244   76146    6681
```

Fig. 4.74

The blocks of digits are "digits" in the base-10^5 system. If we interpret the result as a number in the base-10 system, all four digit blocks must be increased to 5 digits by adding a leading zero. No guarantee can be given for the last digit '1', since we rounded off in line 110. In fact the 250th and 251st decimals are 08. They have been correctly rounded off to 10. In this way D. Gillies and D. Wheeler calculated the number e to one million decimal places (1964). They used the basis $B = 10^{10}$.

Exercises

1. Calculate the number e to a) 500 b) 1000 decimal places.
2. In 1971, J. Dutka calculated $\sqrt{2}$ to 1000000 decimal places. The first 100 decimal places are:

   1·41421 35623 73095 04880 16887 24209 69807 85696 71875 37694

   80731 76679 73799 07324 78462 10703 88503 87534 32764 15727

   Check this by squaring.

   Hint. Regard the blocks of digits as digits of the system with base $B = 10^5$, store the digits in X(0) to X(20) and multiply in the way normally used on paper for multi-digit numbers [long multiplication].

5. COMBINATORICS AND PROBABILITY

5.1 Programs for the Pascal triangle

If a set has n members the number of its subsets which contain exactly s members is denoted by $\binom{n}{s}$. This number $\binom{n}{s} = b(n,s)$ is called a binomial coefficient or a Pascal number.

From the definition it easily follows that

I $\quad \binom{n}{0} = \binom{n}{n} = 1$

II $\quad \binom{n}{s} = \binom{n}{n-s}$

III $\quad \binom{n}{s} = \binom{n-1}{s-1} + \binom{n-1}{s}$

IV $\quad \binom{n}{s} = \frac{n}{s}\binom{n-1}{s-1}$

The recursion IV gives

V $\quad \binom{n}{s} = \frac{n}{s} \cdot \frac{n-1}{s-1} \cdot \frac{n-2}{s-2} \cdot \ldots \cdot \frac{n-s+1}{1}$

In the nth row and the sth column of the table 5.1 is the Pascal number $\binom{n}{s}$. This table is called the Pascal triangle or Pascal matrix. We construct three programs which print this triangle.

n \ s	0	1	2	3	4	5	6	7	8...
0	1								
1	1	1							
2	1	2	1						
3	1	3	3	1					
4	1	4	6	4	1				
5	1	5	10	10	5	1			
6	1	6	15	20	15	6	1		
7	1	7	21	35	35	21	7	1	
8	1	8	28	56	70	56	28	8	1
⋮									

Fig. 5.1 Pascal triangle

a) We change the order of the factors in the numerator of V:

$$\binom{n}{s} = 1 \cdot \frac{n}{1} \cdot \frac{n-1}{2} \cdot \frac{n-2}{3} \cdot \ldots \frac{n-s+1}{s}$$

If this product is evaluated from left to right, the partial products run through the sequence $\binom{n}{0}, \binom{n}{1}, \binom{n}{2}, \ldots \binom{n}{s}$. Thus we obtain the four programs in Figs. 5.2 to 5.5. The first prints $\binom{n}{s}$. The second prints the series $\binom{n}{0}, \binom{n}{1}, \ldots, \binom{n}{s}$. The third prints the nth row of the Pascal triangle. The fourth prints rows 0 to M of the Pascal triangle.

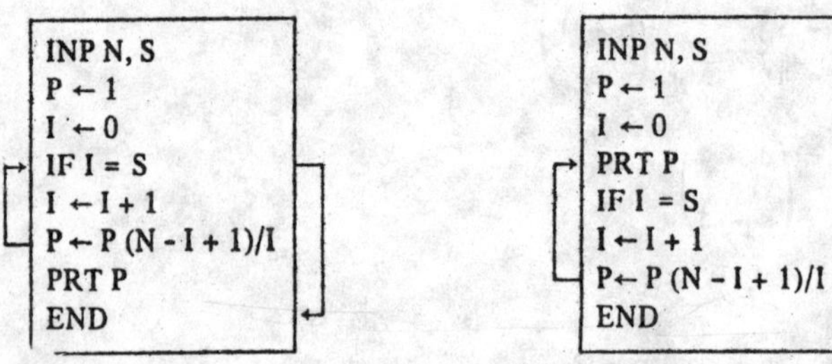

Fig. 5.2 Fig. 5.3

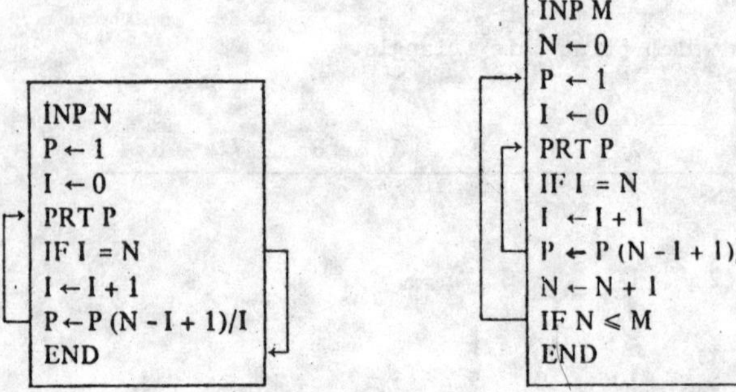

Fig. 5.4 Fig. 5.5

b) Using I and III we can obtain the Pascal triangle by addition. I gives the boundary elements. The inner elements are calculated from III by addition. Fig. 5.6 shows the corresponding BASIC program. The rows 10-70 calculates the Pascal matrix and rows 80 - 130 print it.

```
10 DIM B (15, 15)
20 FOR N = 0 TO 12
30      B (N, 0) = B (N, N) = 1
40      FOR S = 1 TO N - 1
50           B (N, S) = B (N - 1, S - 1) + B (N - 1, S)
60      NEXT S
70 NEXT N
80 FOR N = 0 TO 12
90      FOR S = 0 TO N
100          PRINT B (N, S);
110     NEXT S
120     PRINT
130 NEXT N
140 END
```

```
1
1 1
1 2 1
1 3 3 1
1 4 6 4 1
1 5 10 10 5 1
1 6 15 20 15 6 1
1 7 21 35 35 21 7 1
1 8 28 56 70 56 28 8 1
1 9 36 84 126 126 84 36 9 1
1 10 45 120 210 252 210 120 45 10 1
1 11 55 165 330 462 462 330 165 55 11 1
1 12 66 220 495 792 924 792 495 220 66 12 1
```

Fig. 5.6

c) A few micro-computers do not allow doubly-indexed variables. So we now write an interesting program which uses only the store locations R(0), R(1), R(2),.... . Consider one row of the Pascal triangle, e.g. the fifth: 1, 5, 10, 5, 1. The next row, i.e. the sixth, is obtained by the following calculation

```
     1   5   10   10    5    1
+        1    5   10   10    5    1
---------------------------------------
     1   6   15   20   15    6    1
```

Fig. 5.7 shows a program which uses this property. When N is input it prints rows 0 to N of the Pascal triangle.

```
10 DIM R (20)
20 INPUT N
30 FOR  J = 0 TO N
40        A = 0
50        B = 1
60        FOR I = 0 TO J
70              R (I) = A + B
80              PRINT R (I);
90              A = B
100             B = R (I + 1)
110       NEXT I
120       PRINT
130 NEXT J
140 END
```

```
1
1 1
1 2 1
1 3 3 1
1 4 6 4 1
1 5 10 10 5 1
1 6 15 20 15 6 1
1 7 21 35 35 21 7 1
1 8 28 56 70 56 28 8 1
1 9 36 84 126 126 84 36 9 1
1 10 45 120 210 252 210 120 45 10 1
1 11 55 165 330 462 462 330 165 55 11 1
1 12 66 220 495 792 924 792 495 220 66 12 1
```

Fig. 5.7

Exercises

1. Translate Figs. 5.2 to 5.5 into BASIC and run them.

2. a) Re-write Fig. 5.7 so that when m is input it prints the Pascal triangle modulo m.

   b) Let m = 2. Which rows contain nothing but ones? Which rows contain nothing but zeros, apart from the two ones at the edge?

5.2 Frequency-tables

1st Example: a dice-problem

A dice is rolled 4 times. The outcome is a 4 tuple (X, Y, Z, U). There are $6^4 = 1296$ possible outcomes, which we classify according to the sum of the spots $S = X + Y + Z + U$. How often do the sums 4, 5, . . . 24 occur?

```
10 DIM N (25)
20 FOR I = 4 TO 24
30      N (I) = 0
40 NEXT I
50 FOR X = 1 TO 6
60      FOR Y = 1 TO 6
70           FOR Z = 1 TO 6
80                FOR U = 1 TO 6
90                     S = X + Y + Z + U
100                    N (S) = N (S) + 1
110               NEXT U
120          NEXT Z
130     NEXT Y
140 NEXT X
150 FOR I = 4 TO 24
160     PRINT N (I);
170 NEXT I
180 END

1 4 10 20 35 56 80 104 125 140 146 140 125 104 80 56 35 20 10 4 1
```

Fig. 5.8

	X	Y	Z	U
a)	1	1	1	1
b)	1	1	1	6
c)	1	1	6	6
d)	1	6	6	6
e)	6	6	6	6

Fig. 5.9

The program in Fig. 5.8 solves the problem. Here are some comments on it:

10 reserves locations denoted by N(4) to N(24). In N(S) is stored the frequency of the sum S. 20-40 set the frequency counts to 0. 150-170 print these frequency numbers, after each of the 1296 cases has been assigned to the correct counter. 50-140 is the kernel of the program. It consists of four nested loops and works like the mileometer of a car, but only with the digits 1 to 6 instead of 0 to 9. Initially, X = Y = Z = U = 1 is set, and we have the situation (a) in Fig. 5.9. The innermost loop runs from 1 to 6. Each time U = 6, Z is increased by 1 and U is set back to 1. After 36 steps we have the situation c). Now Y increases by 1, and Z and U are set back to 1. In situation d) the computer leaves the third loop

and X increases by 1, etc. After 1296 steps the situation e) is reached, the computer leaves the outermost loop and begins to print.

2nd Example: The money-changing problem

In how many ways can £1 be given in change using smaller coins?

[We exclude ½p and 20p] Let these coins consist of X, Y, Z, U, V respectively of 1p, 2p, 5p, 10p and 50p pieces.

Then $X + 2Y + 5Z + 10U + 50V = 100.$

We seek the number W of solutions in this equation in non-negative whole numbers. The program in Fig. 5.10 determines W. We shall not give any commentary. The reader may wish to consider its method of working himself.

```
 10 FOR V = 0 TO 2
 20       FOR U = 0 TO 10 - 5 * V
 30             FOR Z = 0 TO 20 - 10 * V - 2 * U
 40                   FOR Y = 0 TO 50 - 25 * V - 5 * U
 50                         FOR X = 0 TO 100 - 50 * V - 10 * U - 5 * Z - 2 * Y
 60                               IF X + 2 * Y + 5 * Z + 10 * U + 50 * V < > 100 THEN 80
 70                               W = W + 1
 80                         NEXT X
 90                   NEXT Y
100             NEXT Z
110       NEXT U
120 NEXT V
130 PRINT "£1 can be changed in"; W; "ways"
140 END
```

£1 can be changed in 2498 ways

Fig. 5.10

Exercises

1. In the USA there are coins for 1, 5, 10, 25, 50 cents. In how many ways can a dollar be given in smaller coins?

   Note: The exercise is considerably simplified if one notes that the number of 1 cent pieces must be divisible by 5.

2. In Switzerland there are coins for 1, 2, 5, 10, 20, 50 centimes In how many ways can a Swiss franc be given in small change?

3. A dice is thrown 6 times. How often does the sum of scores take the value 21, out of the 6^6 possibilities? In other words: How many solutions has the equation $X + Y + Z + U + V + W = 21$ where each variable may take values from 1, 2, . . . 6?

5.3 Permutations

a) Permutations as orderings

We take three different objects and call them 1, 2, 3. An ordering of these objects in a row is called a permutation of the objects. There are six permutations of 1, 2, 3:

123, 132, 213, 231, 312, 321.

As is well-known, there are $1.2.3.\ldots.n = n!$ (Read as "n factorial") permutations of n objects. Also 0! is defined as 1. In Exercise 1 we shall test three approximate formulae for n! But before that we need to introduce a new symbol. A complicated function f is often approximated for large n by a simpler function g. If

$$\frac{f(n)}{g(n)} \longrightarrow 1 \quad \text{as } n \to \infty ,$$

then we write $f(n) \sim g(n)$ (read as $f(n)$ is asymptotically equal to $g(n)$).

The following approximate formulae will not be proved here:

(1) $$n! \sim \sqrt{2\pi n}\,\left(\frac{n}{e}\right)^n$$ (James Stirling 1730)

(2) $$n! \sim \sqrt{2\pi n}\,\left(\frac{n}{e}\right)^n \left(1 + \frac{1}{12n}\right)$$

(3) $$n! \sim \sqrt{2\pi n}\,\left(\frac{n}{e}\right)^n \left(1 + \frac{1}{12n} + \frac{1}{288n^2}\right)$$

b) Generation of a random permutation

We wish to instruct the computer to print the elements of the set $\{1, 2, \ldots n\}$ in random order, so that all n! permutations are equally likely. This is an important, interesting and non-trivial problem. We give an elegant solution.

First the numbers 1 to n are stored in the locations L(1) to L(n). A number is then chosen randomly from 1 to n - suppose it is k. Then L(k) is interchanged with L(n) and n is reduced by 1. The last number in the list L is now in its final place. This step is repeated until n = 1. (Fig. 5.11)

```
I ← 1
L (I) ← I
I ← I + 1
IF I ⩽ N
K ← [N · RND] + 1
C ← L (N)
L (N) ← L (K)
L (K) ← C
N ← N − 1
IF N ⩾ 2
END
```

Fig. 5.11

program in Fig. 5.12 prints m permutations of the set $\{1, 2, . . . n\}$.

```
10 READ N, M
20 DIM L (N)
30 FOR J = 1 TO M
40      FOR I = 1 TO N
50          L (I) = I
60      NEXT I
70      FOR I = N TO 2 STEP − 1
80          K = INT (I * RND) + 1
90          C = L (I)
100         L (I) = L (K)
110         L (K) = C
120     NEXT I
130     FOR I = 1 TO N
140         PRINT R (I);
150     NEXT I
160     PRINT
170 NEXT J
180 DATA 10, 10
190 END
```

```
 9   8   1   3  10   7   5   4   6   2
 4   1   6   9  10   7   8   3   5   2
10   4   8   1   7   6   5   3   9   2
10   9   5   1   3   8   7   2   6   4
 6   8   5   7   4  10   9   3   2   1
 9   7   1   5   3   2   6  10   8   4
10   4   8   7   1   5   9   3   6   2
10   8   7   1   2   4   5   3   9   6
 1   9   5   2  10   8   4   3   7   6
 2   5   6  10   9   8   7   3   4   1
```

Fig. 5.12

Exercises

1. a) We denote the right hand sides of (1), (2), (3) by $f(n)$, $g(n)$, $h(n)$.

   Print a table of

$$\frac{n!}{f(n)}, \quad \frac{n!}{g(n)}, \quad \frac{n!}{h(n)} \qquad \text{for } n = 1, 2, \ldots 10, 20, 30, 40, 50.$$

   b) Determine the quotients in (a) also for $n = 100$. How is over-flow to be avoided?

   c) Permutations as re-orderings

In the last section permuations were regarded as orderings of objects. Now we want to think of them as re-orderings or functions.

e.g. the permutation

$$7 \quad 10 \quad 8 \quad 1 \quad 5 \quad 9 \quad 2 \quad 4 \quad 3 \quad 6$$

can be thought of as the function

$$f = \begin{pmatrix} 1 & 2 & 3 & 4 & 5 & 6 & 7 & 8 & 9 & 10 \\ 7 & 10 & 8 & 1 & 5 & 9 & 2 & 4 & 3 & 6 \end{pmatrix}$$

A permutation $f = \begin{pmatrix} 1 & 2 & & n \\ f(1) & f(2) & \ldots & f(n) \end{pmatrix}$ can be represanted as a graph as follows: In the plane n points are chosen and for each $i \in \{1, 2, 3, \ldots n\}$ an arrow is drawn from i to $f(i)$.

Let

$$f_1 = \begin{pmatrix} 1 & 2 & 3 & 4 & 5 & 6 & 7 & 8 & 9 & 10 \\ 5 & 4 & 6 & 9 & 7 & 8 & 10 & 2 & 3 & 1 \end{pmatrix}$$

$$f_2 = \begin{pmatrix} 1 & 2 & 3 & 4 & 5 & 6 & 7 & 8 & 9 & 10 \\ 3 & 4 & 5 & 1 & 2 & 9 & 7 & 8 & 10 & 6 \end{pmatrix}$$

$$f_3 = \begin{pmatrix} 1 & 2 & 3 & 4 & 5 & 6 & 7 & 8 & 9 & 10 \\ 3 & 9 & 2 & 1 & 6 & 5 & 4 & 10 & 7 & 8 \end{pmatrix}.$$

$$f_4 = \begin{pmatrix} 1\ 2 & 2 & 3 & 4 & 5 & 6 & 7 & 8 & 9 & 10 \\ 2 & 44 & 8 & 3 & 10 & 9 & 1 & 6 & 5 & 7 \end{pmatrix}$$

Fig. 5.13 shows their graphs. They break up into so-called cycles. Under each graph the permutation is given in so-called cycle-notation. Fixed-points (these are points with $f(i) = i$) give unicycles, which are usually omitted.

The permuation f_4 is called cyclic, since it consists of one cycle.

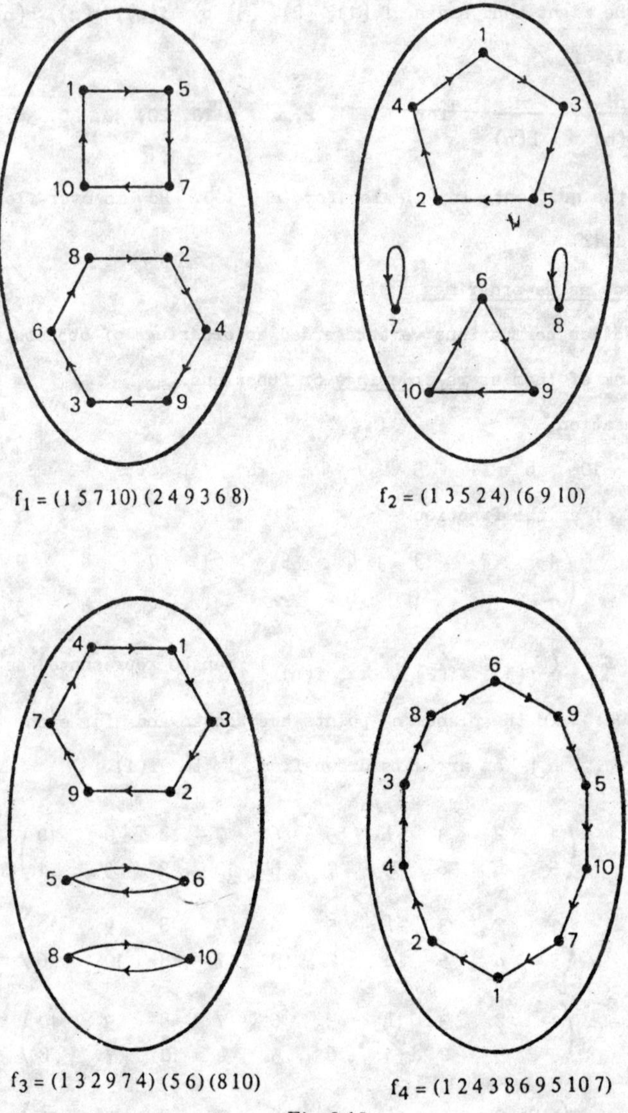

$f_1 = (1\ 5\ 7\ 10)\ (2\ 4\ 9\ 3\ 6\ 8)$

$f_2 = (1\ 3\ 5\ 2\ 4)\ (6\ 9\ 10)$

$f_3 = (1\ 3\ 2\ 9\ 7\ 4)\ (5\ 6)\ (8\ 10)$

$f_4 = (1\ 2\ 4\ 3\ 8\ 6\ 9\ 5\ 10\ 7)$

Fig. 5.13

(d) Permutation programs

Random permutations are excellent number material, whose production leads to instructive programs.

1st Example: the inverse permutation

If, in the graph of a permuation f, all the arrows are reversed, one obtains the inverse permutation f^{-1}. Successive performance of f and f^{-1} gives the identical permutation I :

$$f \circ f^{-1} = f^{-1} \circ f = I = \begin{pmatrix} 1 & 2 & \ldots & n \\ 1 & 2 & \ldots & n \end{pmatrix}$$

Let X(1), X(2), . . ., X(n) be a permutation of $\{1, 2, \ldots n\}$. The inverse permutation Y(1), Y(2), . . . Y(n) is obtained by carrying out the assignment Y(X(i))← i for i = 1 to n.

2nd Example: the order of a permutation

Let f be a permutation of $\{1, 2, \ldots, n\}$ and I be the identical permutation. The smallest natural number p such that $f^p = f \circ f \circ f \circ \ldots \circ f = I$ is called the order or the period of f. We write a program which determines the order of X(1), X(2), . . . X(n). (Fig.5.14).

```
10 P = 1
20 FOR I = 1 TO N        Y (I) = X (I)       NEXT I
30 P = P + 1
40 FOR I = 1 TO N        Y (I) = X (Y (I))   NEXT I
50 FOR I = 1 TO N
60        IF Y (I) ≠ I THEN 30
70 NEXT I
80 PRINT P
90 END
```

Fig. 5.14

It is dangerous to run this program for large n, since the order can be very large. It is better to determine the order via the cycle-lengths. For example, the permutation shown in Fig.5.15 f = (1 2 3 4 5)(6 7 8)(9 10) has the period p = 30. The period p is clearly the highest common factor of the cycle-lengths.

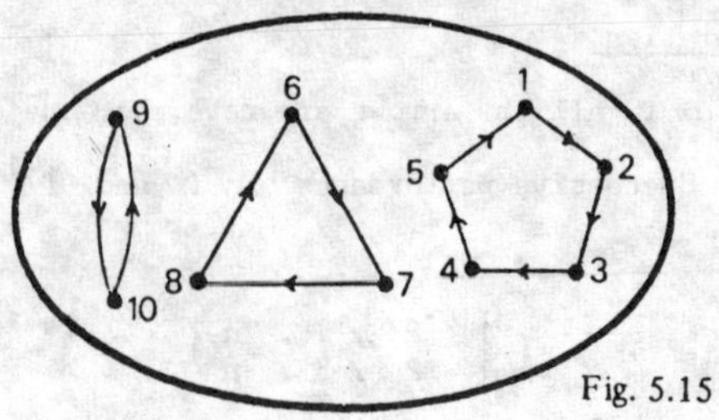

Fig. 5.15

3rd Example: decomposition of a permutation into cycles

Let a permutation of $\{1, 2, \ldots, n\}$ be stored in L(1), L(2), ... L(n). It is to be decomposed into cycles. We start with I = 1 and store this first element in E (Fig.5.16). Then we repeat the assignment I ← L(I) until E reappears, with which the cycle is closed. The complete sequence of values from E onwards is printed out. Each such element is marked so that one knows it has already been dealt with. The simplest way of marking is to use L(I) ← – L(I). After printing, but before marking, we must make a copy K of I, for the iteration I ← L(I) destroys the element I. We need it, however, for the marking L(K) ← – L(K). Since we begin with 1, this element is always treated as marked i.e. the unmarked elements are such that L(I) > 1. If a cycle is completed we look (rows 90-110) for the first unmarked element and with that start a new cycle. If all elements are marked we have finished. Because of line 80, the cycles are printed on separate lines.

```
 10 I = 1
 20 E = I
 30      PRINT I;
 40      K = I
 50      I = L (I)
 60      L (K) = – L (K)
 70      IF I < > E THEN 30
 80 PRINT
 90 FOR I = 1 TO N
100      IF L (I) > 1 THEN 20
110 NEXT I
120 END
```

Fig. 5.16

Exercises

2. Write a program which creates a permutation of $\{1, 2, . . ., 100\}$ and prints out its cycles.
3. Write a program which creates a permutation of $\{1, 2, . . ., 100\}$ and prints out the number of cycles.
4. Repeat the program in Exercise 3 50 times and so estimate the average number of cycles in a permutation of $\{1, 2, . . . , 100\}$. It can be shown that the average value is $1 + \frac{1}{2} + \frac{1}{3} + \; . . . \frac{1}{100}$ (See [5]).

4th Example: The Josephus-Permutation

During the Jewish rebellion against Rome (AD 70), 40 Jews were shut in a cave. In order to avoid slavery, they agreed upon a program of mutual destruction. They would stand in a circle and number themselves from 1 to 40. Then every seventh person was to be killed until only one was left who would commit suicide.

Flavius Josephus, who later wrote the story, so arranged matters that he was the one who was left. Of course, he didn't carry out the last step.

This story is the source of the Josephus-problem: n people are placed in a circle and numbered from 1 to n. Starting with number m, every m^{th} person is removed, whereupon the circle is closed up. The aim is to find the sequence $J_{n,m}$ of those removed. e.g. for n = 8, m = 3 we have

$$J_{8,3} = 3\ 6\ 1\ 5\ 2\ 8\ 4\ 7 \text{ or displayed, } J_{8,3} = \begin{pmatrix} 1 & 2 & 3 & 4 & 5 & 6 & 7 & 8 \\ 3 & 6 & 1 & 5 & 2 & 8 & 4 & 7 \end{pmatrix}$$

Fig. 5.17 shows a program which, when m and n are input, outputs the so-called Josephus-permutation $J_{n,m}$. Initially the numbers 1 to n are stored in L(1) to L(n). H counts those removed. X counts the places from 1 to n and then begins again at 1. Y counts the individuals from 1 to m. If Y = m, then that individual is removed, L(X) is set to 0 and Y also reset to 0. The input values were m = 7, n = 40. So Josephus chose place number 24.

```
10 DIM L (100)
20 INPUT M, N
30 FOR I = 1 TO N    L (I) = I   NEXT I
40 H = 0; X = Y = 1
50 X = X + 1
60 IF X > N THEN X = X - N
70 IF L (X) = 0 THEN 50 ELSE Y = Y + 1
80 IF Y < M THEN 50 ELSE PRINT X;
90 H = H + 1; Y = L (X) = 0
100 IF H < N THEN 50
110 END
```

Fig. 5.17

7 14 21 28 35 2 10 18 26 34 3 12 22 31 40 11 23 33 5 17 30 4 19 36 9 27 6 25 8 32 16 1 38 37 39 15 29 13 20 24

Exercises

5. Fig. 5.17 shows a very advanced BASIC-version. Translate the program into the dialect of your machine and verify that the following permuations are cyclic:

a) $J_{n,2}$ for n = 2, 5, 6, 9, 14, 18 b) $J_{n,3}$ for n = 3, 5, 27

c) $J_{n,4}$ for n = 5, 10 d) $J_{n,7}$ for n = 11, 21, 35.

5.4 Probability problems

The solution of many probability problems require a large amount of calculation. We consider two typical examples.

1st Example: The Birthday problem (see [4]).

In a room there are n people. What is the probability that at least two people have the same birthday?

It is easier to determine the probability $q_n = 1 - p_n$, that they all have different birthdays. In the language of urns and balls the problem becomes: n balls are distributed randomly in turn into 365 urns. What is the probability q_n that each ball is placed in an empty urn?

The 1st, 2nd, 3rd.... n^{th} balls fall in an empty urn with probability

$$\frac{365}{365}, \frac{364}{365}, \frac{363}{365}, \ldots, \frac{365-n+1}{365}$$

By multiplication we have

$$q_n = \frac{365}{365} \cdot \frac{364}{365} \cdots \frac{365-n+1}{365}, \text{ and } p_n = 1 - q_n.$$

Fig. 5.18 prints the pairs (i, p_i) for $i = 1$ to n using the recursion

$$q_1 = 1, \quad q_{n+1} = q_n \cdot \frac{365 - n}{365} .$$

```
10 INPUT N
20 Q = 1
30 FOR I = 1 TO N
40        PRINT I, 1 – Q
50        Q = Q* (365 – I)/365
60 NEXT I
70 END
```

Fig. 5.18

Exercises

1. Run the program in Fig. 5.18 for $n = 23$, with a computer or a pocket calculator. The result is surprising.
2. Alter the program of Fig. 5.18 so that the pair (n, p_n) is printed as soon as $p_n > s$, and run it for

   a) $s = 0.9$ b) $s = 0.99$ c) $s = 0.999$ d) $s = 0.9999$
3. On the planet X, a year has 1000 days, It is required to find the probability that among n people at least two individuals have the same birthday. What is the smallest n for which

   a) $p_n > \frac{1}{2}$ b) $p_n > 0.9$ c) $p_n > 0.99$ d) $p_n > 0.999$.
4. On the planet Y the years has x days; p_n has the same meaning as in exercise 3. It is known that $n = 50$ gives the smallest value of n for which $p_n > \frac{1}{2}$. What are the possible values of x ?

2nd Example: The binomial distribution

In Fig. 5.19 let p and $q = 1 - p$ be the probabilities of the outcome 1 (success) and 0 (failure). If this spinner is spun n times, then

$$b(x) = \binom{n}{x} p^x q^{n-x} , \quad x = 0, 1, 2, \ldots , n$$

is the probability of exactly x successes. More exactly, one should write $b(x; n, p)$ in order to make clear the dependence on the parameters n and p. The function b is called the binomial distribution. This very frequently used distribution is awkward to calculate. Thus it is a rewarding subject for programming exercises. Fig. 5.20 prints the pair (x, b(x)) for values of x from 0 to n when n and p are input. The program depends on the easily-verified recursion:

$$b(0) = q^n \ , \quad b(x) = \frac{n - x + 1}{x} \cdot \frac{p}{q} . b(x - 1) \ , \quad x = 1, 2, 3, \ldots, n.$$

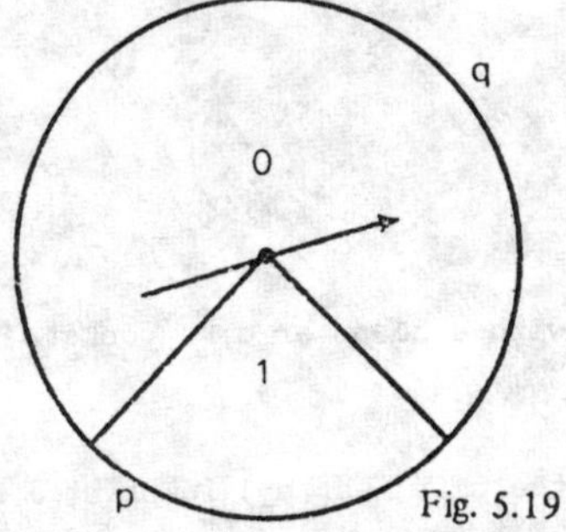

Fig. 5.19

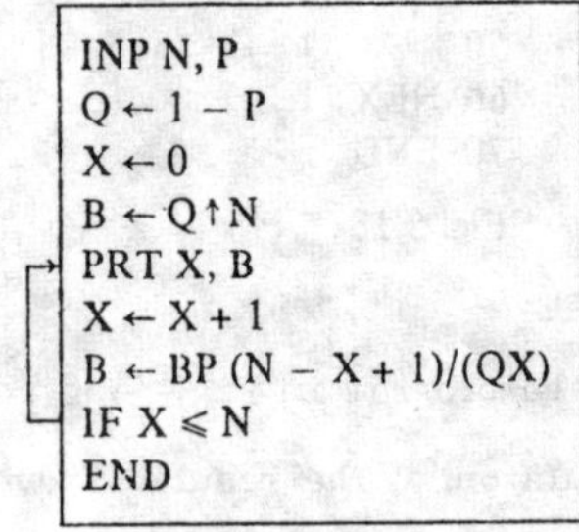

Fig. 5.20

In practice we usually want to find the sums

b(0) + b(1) + . . . + b(a) (Probability of at most <u>a</u> successes)

b(a) + b(a + 1) + . . . + b(n) (Probability of at least <u>a</u> successes)

b(c) + b(c + a) + . . . + b(d) (Probability of between c and d successes, inclusive).

<u>Remark</u> The program in Fig. 5.20 has one weakness. The assignment B← Q ↑ N can lead to "underflow". For example, for n = 330 and $q = \frac{1}{2}$, my computer 'goes on strike', since $\frac{1}{2^{330}} < 10^{-99}$.

<u>Exercises</u>

5. Write a program which, when a, n and p are input, outputs the sum b(0) + b(1) + . . . + b(a). Notice that the same program can be used to calculate b(a) + b(a + 1) + . . . + b(n) if one inputs the triple (n - a, n, q) instead of (a, n, p).

6. In 100 throws of a good coin let 'tails' appear x times. How large is the probability that a) $x < 40$ b) $|x - 50| > 10$? (Use the program of Exercise 5).

7. 600 throws of a dice give 120 sixes (successes). How large is the probability that when an unbiased dice is thrown deviations from the expected number (100) will occur which are as large as, or larger than, this? Hint. Here $n = 600$, $p = \frac{1}{6}$. We require the sum

$$b(0) + b(1) + \ldots + b(80) + b(120) + b(121) + \ldots + b(600).$$

[† An additional note provided by Prof. Engel]: Underflow can be avoided by taking logarithms. The program in Fig. 5.23 computes $b(C) + \ldots + b(D)$; a left tail can be summed by setting $C = 0$, and a right tail by setting $D = N$.

3rd Example: Finding an asymptotic formula

The probability of exactly n tails in 2n tosses of a true coin amounts to

$$b_n = \binom{2n}{n} 2^{-2n} = \frac{1.3.5. \ldots (2n-1)}{2.4.6. \ldots 2n}$$

It can be shown, but only with great labour, that

$$b_n \sim \frac{1}{\sqrt{\pi n}}$$

(see [5]). This formula can be verified empirically with the computer and even improved. The program in Fig. 5.21 prints table 5.22. It can easily be checked that $\sqrt{\pi n}.\, b_n$ tends to 1 in such a way that doubling n halves the error in $\sqrt{\pi n} b_n$. This implies that $\sqrt{\pi n}.\, b_n \sim 1 + \frac{a}{n}$.

For $n = 2000$ this gives

$$1 + \frac{a}{2000} \simeq 0{\cdot}9999374946$$

$$\text{or} \quad a \doteq -0{\cdot}125108 \doteq -\frac{1}{8}$$

$$\text{i.e.} \quad b_n \sim \frac{1}{\sqrt{\pi n}} \left(1 - \frac{1}{8n}\right)$$

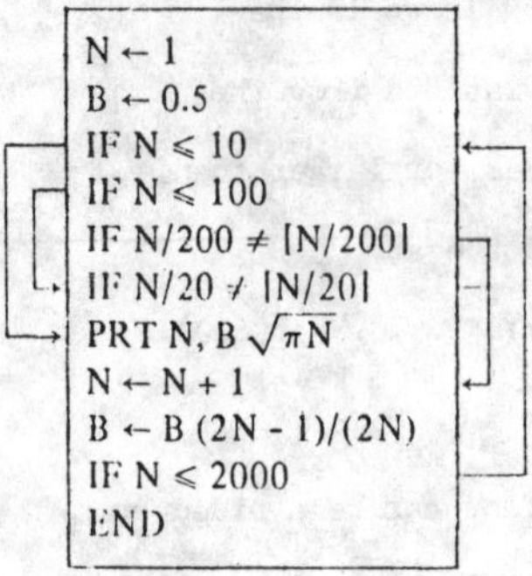

```
   N ← 1
   B ← 0.5
┌─ IF N ≤ 10                  ←─┐
│┌ IF N ≤ 100                    │
││ IF N/200 ≠ [N/200]  ─┐        │
│└→ IF N/20 ≠ [N/20]   ─┤        │
└→ PRT N, B √(πN)                │
   N ← N + 1            ←┘        │
   B ← B (2N - 1)/(2N)           │
   IF N ≤ 2000         ──────────┘
   END
```

Fig. 5.21

n	$\sqrt{\pi n}\ b_n$	n	$\sqrt{\pi n}\ b_n$
1	0.8862269255	80	0.9984387300
2	0.9399856030	100	0.9987507859
3	0.9593687887	200	0.9993751956
4	0.9693106997	400	0.9996875482
5	0.9753500771	600	0.9997916871
6	0.9794056043	800	0.9998437602
7	0.9823161772	1000	0.9998750050
8	0.9845064055	1200	0.9998958351
9	0.9862141369	1400	0.9999107137
10	0.9875829288	1600	0.9999218726
20	0.9937701371	1800	0.9999305516
40	0.9968799587	2000	0.9999374946
60	0.9979188592		

Fig. 5.22

Exercises

8. For a) $n = 100$ and $n = 200$ b) $n = 1000$ and $n = 2000$ verify that $1 - \sqrt{\pi n}.\ b_n$ is almost exactly halved when n is doubled.

9. a) A table for the sequence $c_n = \sqrt{\pi n}.\ b_n \Big/ (1 - 1/8n)$ is to be printed, like that in Fig. 5.22

   b) Verify that c_n converges such that the error is divided by 4 when n is doubled, i.e. that $c_n \sim 1 + \dfrac{b}{n^2}$.

   c) Determine b for $n = 200$, and show this value can be expressed approximately in the form $b = \dfrac{1}{m}$, where m is a whole number.

```
 10  INPUT N,P,C,D
 20  Q  = 1 - P : R = LOG(P/Q) : L = N*LOG(Q )
 30  FOR  X = 1  TO  C                          } computes
 40      L = L + R + LOG( (N-X+1) /X)           } ln b(c)
 50  NEXT X
 60  S = EXP(L)                         computes  b(c)
 70  FOR  X = C + 1  TO D                      } computes
 80      L = L + R + LOG( (N-X+1) /X)          } b(c) + . . . + b(d)
 90      S = S + EXP(L)                        }
100  NEXT X
110  PRINT S
120  END
```

Fig. 5.23

6. SIMULATION OF RANDOM PROCESSES

6.1 The random number generator

The random number generator was introduced in 1.3.6 and often used after that. Let N be a natural number. Then we already know the following BASIC commands

a) RND chooses a real number at random in the interval (0, 1)

b) N*RND chooses a real number at random in the interval (0, N)

c) INT (N*RND) chooses a random element of the set $\{0, 1, 2, \ldots N-1\}$

d) INT (N*RND) + 1 chooses a random element of the set $\{1, 2, 3, \ldots N\}$

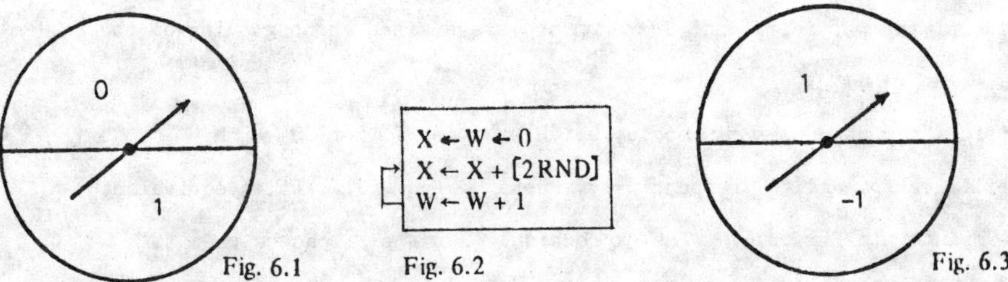

Fig. 6.1 Fig. 6.2 Fig. 6.3

If we mark the faces of a true coin with 0 (failure) and 1 (success), then each toss chooses randomly an element of the set $\{0, 1\}$. The same thing occurs when the spinner in Fig. 6.1 is used. The toss of a coin is easily simulated. Each of the expressions

$$[2 * \text{RND}] \quad , \quad [\text{RND} + 0.5]$$

can be used to simulate the toss of a fair coin. Fig. 6.2 simulates a sequence of coin tosses, in which X and W count the '1''s and the throws. Sometimes it is appropriate to mark the side of a coin with +1 and -1. A fair coin is then equivalent to the spinner in Fig. 6.3. With this we can simulate an important random process, which will now be described.

A particle moves from point to point ($0, \pm 1, \pm 2, \ldots$) on the x axis, so that each second it takes a step of length 1 to the left or the right, each with probability $\frac{1}{2}$. This movement is called <u>a symmetrical random walk along a straight line</u>. The points $0, \pm 1, \pm 2, \ldots$ are called states.

Each of the expressions :

$$2*\lceil 2\cdot \text{RND} - 1\rceil, \quad 2*\lfloor \text{RND} + 0{\cdot}5\rfloor -1$$

can be used to give one step in a symmetrical random walk on the line.

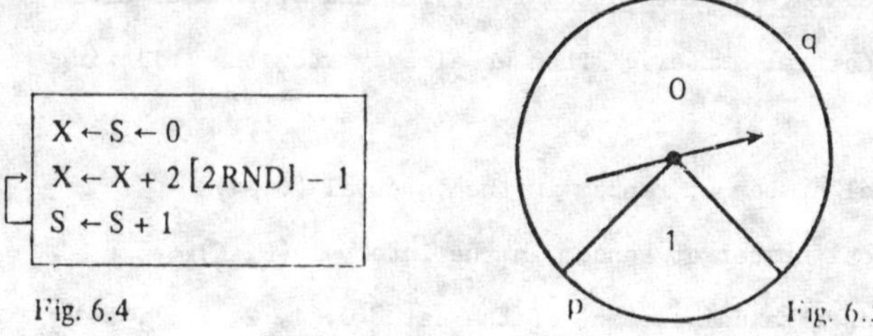

Fig. 6.4 Fig. 6.5

The small program in Fig. 6.4 starts a random walk at 0. X gives the position of the particle, and S counts the steps since the beginning of the walk.

We consider now an asymmetrical coin, which produces 1 or 0 with probabilities p and q respectively, where $p + q = 1$. It is equivalent to the spinner in Fig. 6.5. The expression $\lfloor \text{RND} + p\rfloor$ can be used to simulate one toss. This can be seen with the help of Fig. 6.6. A coin which gives +1 or −1 with probabilities p and q respectively is equivalent to the spinner in Fig. 6.7. With this random apparatus we can simulate the <u>asymmetrical random walk</u> on a line, in which each second the particle moves one step to right or left with probabilities p and q respectively. The expression $2\lfloor \text{RND} + p\rfloor - 1$ can be used to simulate one step of the random walk.

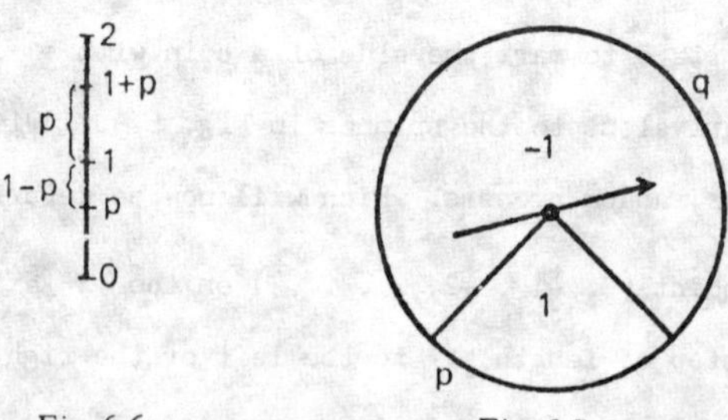

Fig. 6.6 Fig. 6.7

Each random process can easily be simulated with the random number generator. We will learn the art of simulation with the help of numerical examples.

At the same time we shall also make a substantial step forward in the art of programming, for simulation programs are especially instructive from the standpoint of data-processing.

6.2 Simulations with the random number generator

1st Example: Drawing a random sample

In statistics it is often necessary to draw a sample of s members from a set of n elements, (the so-called population). It is hoped to draw conclusions about the population by examining the sample. That is only possible when we are concerned with a random sample. In a random sample, every element of the population has the same probability, $\frac{s}{n}$, of belonging to the sample. A random sample is chosen by numbering each of the elements of the population and choosing s different numbers at random from the numbers allocated.

We need to note which numbers are chosen so that no number is chosen twice. We set up a list L, in which L(I) represents the number I. Then we choose, at random, an element I of the list. If I is chosen, we put L(I) = -1. The program in Fig. 6.8 chooses 6 elements from $\{1, 2, ... 49\}$. (Lotto or Bingo)

```
10 READ N, S
20 DIM L (N)
30 FOR I = 1 TO N
40      L (I) = I
50 NEXT I
60 FOR K = 1 TO S
70      I = INT (N*RND) + 1
80      IF L (I) < 0 THEN 70
90      PRINT I;
100     L (I) = - 1
110 NEXT K
120 DATA 49,6
130 END
```

```
12 20 38 31 22 35
44 14 30  2 31 18
 2 28 19 16 44 24
23 38 42 39 35 48
29 33 17 45  3 21
45 15 26 29  3 30
13 37 10 16 27 15
39 41 15  4 40 19
19 29 37 11 18 31
40 14 20  1 33 10
```

Fig. 6.8

Exercise

1. Neighbours in Lotto. Abel and Cain choose a sample of 6 from 1, 2, ... 49. If the sample contains neighbours (e.g. 18 and 19) then Cain wins. Otherwise Abel wins. The game is to be played 100 (1000) times and Abel's chance of winning estimated. Write a suitable program.

   Hint. Neighbours exist if L(I) + L(I + 1) = -2 is satisfied for some I.

2nd Example: Waiting for a complete set

a) I have thrown a die until all six numbers have appeared. The result was the digit sequence 4 31232 13311 22265. How many throws are needed, on average, to obtain a complete set of 6 digits? (Collector's problem)

b) TOP is a popular make of chocolate in Ruritania. Every block has a token with one of the equally-probable numbers 1, 2,. . . n.

A prize is given for a complete set of numbers. Let T_n be the waiting time for a complete set. We wish to estimate the average value $E(T_n)$ by means of a numerical experiment. First we write a program which collects a complete set from $\{1, 2, . . . n\}$ and prints the waiting time.

We use the following variables:

T counts the number of times a choice has been made.

V counts the number of different collected elements

R is a randomly chosen element of $\{1, 2, . . . , n\}$.

We put L(R) = 1 if R appears for the first time. As soon as V = N, we have a complete set. Fig. 6.9 shows the resulting program.

```
  5 INPUT N
 10 DIM L (N)
 20 T = 0
 30 FOR I = 1 TO N
 40      L (I) = 0
 50 NEXT I
 60 FOR V = 1 TO N
 70      R = INT (N*RND) + 1
 80      T = T + 1
 90      IF L (R) = 1 THEN 70
100      L (R) = 1
110 NEXT V
120 PRINT T
130 END
```

Fig. 6.9

Exercises

2. Modify the program in Fig. 6.9 so that it collects 10 complete sets and prints the 10 waiting times. Run it on the computer for n = 6 (dice) and n = 10 (random decimal digits).

3. Modify the program so that it collects 100 complete sets and only prints the average value of the 100 waiting times. Run it on the computer for n = 6 and n = 10.

3rd Example: The game of craps

Craps is the quickest and most popular American dice game. The rules of the game are:

1. Roll two dice and determine the sum of the scores. S = 7 or S = 11 is an immediate win, S = 2, 3 or 12 is an immediate loss.
2. Any other sum is called the "point" P, and the game is continued until S = 7 (a loss) or S = P (a win) occurs.

```
10 READ N, G, W
20 FOR I = 1 TO N
30     S = INT (6*RND) + INT (6*RND) + 2
40     W = W + 1
50     IF S = 7 OR S = 11 THEN 120
60     IF S = 2 OR S = 3 OR S = 12 THEN 130
70     P = S
80     S = INT (6*RND) + INT (6*RND) + 2
90     W = W + 1
100    IF S <> 7 AND S <> P THEN 80
110    IF S = 7 THEN 130
120    G = G + 1
130 NEXT I
140 PRINT "GAMES = "; N,"WINS = "; G,"LOSSES = ";N - G,"THROWS = ";W
150 DATA 1000, 0, 0
160 END
```

Fig. 6.10

We repeat the game n times and estimate the probability of winning as well as the average game length (number of throws). The expression INT(6*RND) + 1 represents one throw of a die . Thus the sum of the scores for two throws is represented by INT(6*RND) + INT(6*RND) + 2

The variables G and W count the wins and the number of times a pair is thrown. The program in Fig. 6.10 may be understood without further commentary.

Some microcomputers do not allow relations involving AND and OR, of the type seen in lines 50, 60 and 100. These can be written in an alternative way:

```
 50 IF (S - 7) * (S - 11) = 0 THEN 120
 60 IF (S - 2) * (S - 3) * (S - 12) = 0 THEN 130
100 IF (S - 7) * (S - P) ≠ 0 THEN 80
```

Unfortunately this leads to a lot of multiplication. Lines 50 and 60 can be elegantly re-written to avoid multiplication altogether by the following method.

```
50 IF ABS (S - 9) = 2 THEN 120
60 IF ABS (S - 7) ≥ 4 THEN 130
```

For $|S - 9| = 2$ is true precisely when $S = 7$ or $S = 11$. If $|S - 9| \neq 2$ then in particular, $S \neq 11$ and $|S - 7| \geq 4$ is only true for $S = 2$, or 3, or 12.

4th Example: A coin game

Abel says to Cain; "We will toss a coin with sides 0 and 1 until one of the sequences 1111 or 0011 appears. In the first case, you win and in the second, I win. The game is fair, since each sequence has the probability $\frac{1}{16}$".

Is Abel correct? How large is his expectation of winning? What is the average length of a game?

I have repeated the game four times with the result 110110011, 0100000011, 1111, 011011100000011. Abel has won three times and Cain only once. The average game length was 10.75.

The computer is instructed to play n games. W counts the throws, and A Abel's wins. We need only note the last four throws X, Y, Z, U. They make a block of digits which can be thought of as a binary number with value $D = 8X + 4Y + 2Z + U$. The blocks 1111 and 0011 correspond to the values $D = 15$ and $D = 3$ respectively. Conversely, from the value D one can reconstruct the block of digits. For example $D = 13 = 8 + 4 + 1$ corresponds to 1101. The game is at an end as soon as $|D - 9| = 6$. If $|D - 9| \neq 6$, then we put $X = Y$, $Y = Z$, $Z = U$ and ascertain U by the next toss of the coin. This gives the program in Fig. 6.11. For 1000 games, 11986 throws were needed. The average number of throws is about 12. Abel's expectation of winning is approximately 0·741

```
 10 READ N, A, W
 20 FOR I = 1 TO N
 30      X = INT (RND + 0.5)
 40      Y = INT (RND + 0.5)
 50      Z = INT (RND + 0.5)
 60      U = INT (RND + 0.5)
 70      D = 8 * X + 4 * Y + 2 * Z + U
 80      IF ABS (D - 9) = 6 THEN 150
 90      X = Y
100      Y = Z
110      Z = U
120      U = INT (RND + 0.5)
130      W = W + 1
140      GO TO 70
150      IF D <> 3 THEN 170
160      A = A + 1
170 NEXT I
180 PRINT "W= ",W, "A= ",A
190 DATA 1000, 0, 4
200 END
W = 11986      A = 741
```

Fig. 6.11

1 — 2

Fig. 6.12

Exercise:

4. Two players, White and Black, draw alternately. Whoever has the draw chooses one of the numbers 1, 2 at random and places a chip of his own colour on the chosen number in Fig. 6.12. The winner is the one who first has chips on both of the circles.
Write a program which repeats the game n times. Estimate Black's chance of winning and the average game length.

5th Example: Relative frequency and probability

a) We make a sequence of tosses of a coin, and consider the relative frequency of '1' 's in the resulting sequences of 0's and 1's.

$$\frac{E}{N} = \frac{\text{Number of 1's}}{\text{Number of tosses}}$$

By the so-called Strong Law of Large Numbers (see [4]), $\frac{E}{N} \to \frac{1}{2}$ "with probability 1". We wish to follow this 'probability 1' convergence with the computer. The program in Fig. 6.13 tosses a coin 2000 times and prints $\frac{E}{N}$ for N = 10, 20, . . ., 100, 200, . . ., 2000. The table shows that $\frac{E}{N}$ approaches the number $\frac{1}{2}$ in a very irregular way. It can be shown that $\left|\frac{E}{N} - 0{\cdot}5\right| \leq \frac{1}{2\sqrt{N}}$ with probability 0·68 and $\left|\frac{E}{N} - 0{\cdot}5\right| \leq \frac{1}{\sqrt{N}}$ with probability 0·95. The deviation $\sigma = \frac{1}{2\sqrt{N}}$ is called the standard deviation. The deviation $2\sigma = \frac{1}{\sqrt{N}}$ is used as an "alarm signal". If this is overstepped, then there is a strong suspicion that the coin is not fair. The third column in Fig. 6.13 measures the deviations in σ-units. The 2σ boundary is not crossed.

b) Fig. 6.14 shows a decimal spinner. The program in Fig. 6.15 spins it 1000 times and counts the frequency of the outcomes 0 to 9. The observed frequencies B_i differ from the expected frequencies $E_i = 100$. Are these deviations the result of chance or are they so large that they cast doubt on the quality of the 'spinner'? Information on this is given by the quantity

$$\chi^2 = \sum_{i=1}^{n} \frac{(B_i - E_i)^2}{E_i} .$$

It can be shown that $E(\chi^2) = n - 1, \quad \sigma = \sqrt{2(n - 1)}$.

```
10 E = 0
20 FOR N = 1 TO 2000
30      E = E + INT (2 * RND)
40      IF N < = 100 AND N/10 = INT (N/10) THEN 60
50      IF N/100 <> INT (N/100) THEN 70
60      PRINT N, E/N, 2 * SQR (N) * ABS (E/N - 0.5)
70 NEXT N
80 END
```

10	0.3	1.26491
20	0.5	0
30	0.533333	0.36515
40	0.6	1.26491
50	0.56	0.84853
60	0.583333	1.29099
70	0.571429	1.19523
80	0.5625	1.11803
90	0.555556	1.05409
100	0.54	0.8
200	0.525	0.70711
300	0.49	0.34641
400	0.4975	0.1
500	0.504	0.17889
600	0.506667	0.32660
700	0.502857	0.15119
800	0.50625	0.35355
900	0.51	0.6
1000	0.51	0.63246
1100	0.509091	0.60302
1200	0.505	0.34641
1300	0.500769	0.05547
1400	0.496429	0.26726
1500	0.499333	0.05164
1600	0.49375	0.5
1700	0.492353	0.63059
1800	0.493889	0.51854
1900	0.492632	0.64236
2000	0.4945	0.49193

Fig. 6.13

Fig. 6.14

```
10 FOR I = 0 TO 9
20      B (I) = 0
30 NEXT I
40 FOR I = 1 TO 1000
50      D= INT (10*RND)
60      B (D) = B (D) + 1
70 NEXT I
80 FOR I = 0 TO 9
90      PRINT B (I);
100 NEXT I
110 END
```

87 96 102 104 96 106 93 101 110 105

Fig. 6.15

In our case $E(\chi^2) = 9$, $\sigma = \sqrt{18} = 4.24\ldots$. Since our $\chi^2 = 4.32$, is less than the expected value of 9, there is no cause to doubt the quality of the random number generator. The calculation of χ^2 may also be carried out on the computer.

Exercise:

5. Write a program which rolls a dice 600 times, counts the frequencies of the outcomes 1 to 6, and calculates χ^2

6th Example: Runs in binary digit sequences

A fair coin is tossed 20 times with the result 00 11 0000 1 0 11 0000 1 00. There are nine blocks of equal digits, which are called "runs". We have four runs of 1's and five of 0's. A schoolboy was asked to devise 50 tosses of a coin without using any random number equipment. He went about it with great care and submitted the following result:

10101101001011000101100101011000101010011001010100.

At first glance the sequence seems acceptable. It contains 23 ones and 27 zeros. In spite of this it is a poor imitation of randomness. The sequence contains 36 runs. A randomly-generated sequence would hardly ever contain so many runs. We wish to convince ourselves of this by instructing the computer to perform 50 coin tosses 100 times in succession, counting the runs and printing the results each time. The average number of runs is also to be printed. Fig. 6.16 shows the corresponding program. S and T are the last two tosses of the coin. W counts the changes from 0 to 1 or from 1 to 0. The number of runs is one more than the number of changes. The variable R accumulates the number of runs. Row 140 prints the average number of runs $\frac{R}{100}$.

```
10 R = 0
20 FOR J = 1 TO 100
30     W = 0
40     S = INT (2*RND)
50     FOR I = 1 TO 49
60         T = INT (2*RND)
70         W = W + ABS (S - T)
80         S = T
90     NEXT I
100    PRINT W + 1;
110    R = R + W + 1
120 NEXT J
130 PRINT
140 PRINT R/100
150 END
```

Fig. 6.16

```
25 34 16 29 27 27 26 22 22 25 25 26 23 31 29 23 22 22 22 29 28 24 23 26 28
25 23 27 26 19 26 25 23 26 31 30 27 23 29 27 30 23 25 28 28 24 22 24 29 23
21 29 22 24 21 30 27 20 23 24 25 23 29 26 25 26 23 24 22 23 24 27 26 33 27
26 24 26 30 22 28 26 22 25 24 21 27 22 29 27 26 28 31 24 26 25 24 27 24 28
25.43
```

We see that the average number of runs is c.25.5 and that 36 runs or more never occurred in 100 cases. More information can be looked up in [4].

7th Example: Symmetrical random walk. The $\sqrt{n}$ theorem

a) Symmetrical random walk on the line. A particle begins at the origin 0 and each second jumps one step to the left or to the right with probability $\frac{1}{2}$ in each case. Let D_n be the distance from 0 after n steps. The random variable D_n can take the values 0, 2, 4, ... n or 1, 3, 5, ... n according to whether n is even or odd. Let the expected value of D_n^2 be $E(D_n^2)$. We wish to estimate $E(D_n^2)$. For this we simulate m = 100 random walks with n = 10 steps. We print all m values of D_n^2 in order to get an impression of the distribution. The mean of the m values provides an estimate of $E(D_n^2)$. In Fig. 6.17, I counts the random walks and K counts the steps of a walk. X is the current position of the wandering particle, S accumulates the m values of D_n^2. The printed values permit the conjecture that $E(D_n^2) = n$.

```
 10 READ S, N, M
 20 FOR I = 1 TO M
 30     X = 0
 40     FOR K = 1 TO N
 50         X = X + 2*INT (2*RND) - 1
 60     NEXT K
 70     D = X*X
 80     PRINT D;
 90     S = S + D
100 NEXT I
110 PRINT
120 PRINT S/M
130 DATA 0, 10, 100
140 END
```

```
0 4 16 36 16 16 0 4 4 16 16 16 16 0 4 64 0 0 16 4 16 4 16 4 4 16 16 4 4 0 0 0
16 0 4 0 16 4 0 0 0 4 0 4 16 0 4 0 0 4 16 16 4 4 16 4 16 16 16 4 4 4 0 36 4 16
4 4 4 16 4 0 16 4 0 36 4 16 36 16 0 16 4 0 4 0 4 4 64 0 0 16 16 16 4 4 16 4 64 0
9.76
```

Fig. 6.17

b) <u>Symmetrical random walk in the plane</u>. A particle, starting at 0, follows a random path in the plane. At each step it moves with probability $\frac{1}{4}$ one step to left, to right, upwards or downwards. Let D_n be its distance from 0 after n steps. We estimate $E(D_n^2)$. The program in Fig. 6.18 is a slight variation of the previous one. Lines 30 to 80 simulate a random walk of n steps. A is 0 or 1 and B is +1 or −1, each time with probability $\frac{1}{2}$. If A = 0, y is altered by 1 or −1. If A = 1, then x is altered by 1 or −1. This time we are not interested in the individual D_n^2. Five runs of the program gave 8·28, 8·94, 9·74, 10·44, 12·14. These numbers permit the conjecture that $E(D_n^2) = n$ also for the plane.

```
10 READ S, N, M
20 FOR I = 1 TO M
30     X = Y = 0
40     FOR K = 1 TO N
50         A = INT (2*RND)
60         B = 2*INT (2*RND) - 1
70         X = X + A*B
80         Y = Y + (1 - A)*B
90     NEXT K
100    D = X*X + Y*Y
110    S = S + D
120 NEXT I
130 PRINT S/M
140 DATA 0, 10, 100
150 END
```

Fig. 6.18

```
10 READ S, N, M
20 FOR I = 1 TO M
30     X = Y = 0
40     FOR K = 1 TO N
50         A = 2 * PI *RND
60         X = X + COS (A)
70         Y = Y + SIN (A)
80     NEXT K
90     D = X*X + Y*Y
100    S = S + D
110 NEXT I
120 PRINT S/M
130 DATA 0, 10, 100
140 END
```

Fig. 6.19

c) A particle starts at 0 and makes unit steps. The direction of each step is randomly chosen between 0 and 2π. Fig. 6.20 shows 10 steps of this random walk. Once more we wish to estimate $E(D_n^2)$. Fig. 6.21 shows how, from the current position (x, y), the next position is reached. By A ← 2*π*RND, a random angle is chosen in the range 0 to 2π. Then x ← x + cos A, y ← y + sin A gives the next position.

The program in Fig. 6.19 simulates m = 100 random walks with n = 10 steps. Five runs of the program produced the values 10.90, 8.89, 10.40, 9.89, 8.50.

These values again support the conjecture that $E(D_n^2) = n$.

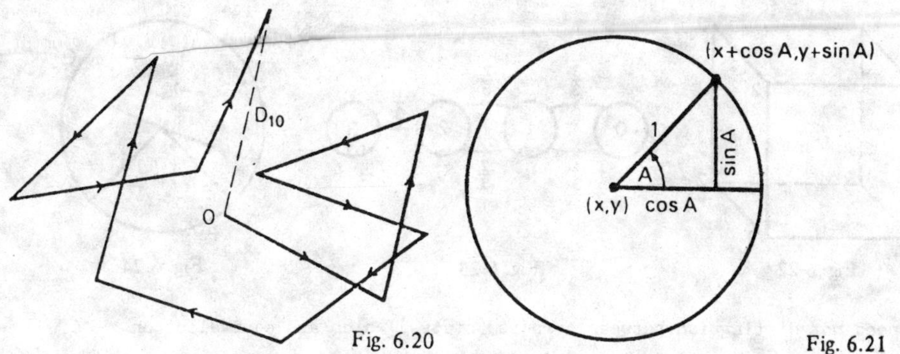

Fig. 6.20

Fig. 6.21

Random walks are discrete approximations to physical diffusion processes. The symmetrical random walk is the discrete version of the Brownian motion. A drop of dye in a thin film of water slowly disperses but retains its roughly circular shape. The single particles of dye move with an irregular motion. The radius of the circle which contains half of the particles, grows with time according to the formula $r(t) = c\sqrt{t}$. We have observed special cases of this result. That is to say, if $E(D_n^2) = n$, then $E(D_n) = c\sqrt{n}$ with $0 < c < 1$. This follows from the in-equality

$$\sqrt{\frac{d_1^2 + d_2^2 + \ldots + d_n^2}{n}} \geqslant \frac{d_1 + d_2 + \ldots + d_n}{n}$$

8th Example: Random walk on a cube

A beetle walks randomly along the eges of a cube. It starts at 0 (Fig.6.22), and takes one minute to walk along one edge. At each corner it chooses one of the three edges, each with probability $\frac{1}{3}$. At corner 3, it stops.

a) Simulate a random walk and print the time it takes (duration), I.

b) Simulate N random walks, print the N times and the average times.

c) Simulate N random walks, print a frequency table for the times and the average time.

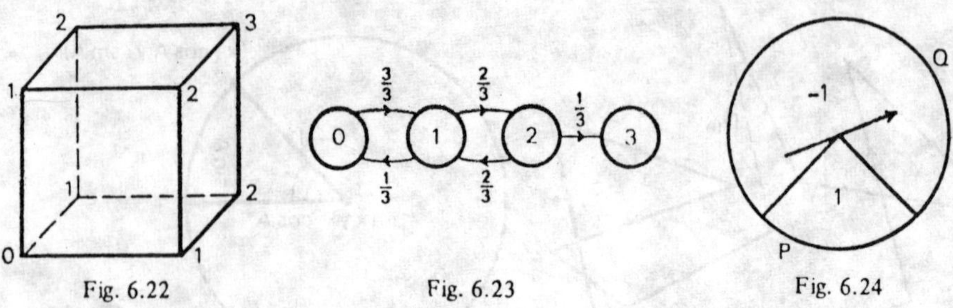

Fig. 6.22 Fig. 6.23 Fig. 6.24

We need not distinguish between corners (states) which are equivalent on grounds of symmetry. The random walk on the cube then reduces to a random walk on Fig. 6.23. We recall Fig. 6.24 which can be used to produce a random walk on a line. If we go from the current state Z to Z + 1 and Z − 1 with probabilities P and Q respectively, we obtain the next state by the assignment

$$Z \leftarrow Z + 2*\left[RND + P\right] - 1.$$

For $Z \lesssim 2$ in Fig. 6.23 we have $P = \frac{3 - Z}{3}$ i.e. for $Z \leqslant 2$,

$$Z \leftarrow Z + 2*\left[RND + \frac{3 - Z}{3}\right] - 1$$

gives one step of the random walk. The program in Fig. 6.25 solves part a).

```
10 I = Z = 0
20 Z = Z + 2 * INT (RND + (3 – Z)/3) – 1
30 I = I + 1
40 IF Z < 3 THEN 20
50 PRINT I
60 END
```

Fig. 6.25

```
10 READ S, N
20 FOR K = 1 TO N
30       I = Z = 0
40       Z = Z + 2 * INT (RND + (3 – Z)/3) – 1
50       I = I + 1
60       IF Z < 3 THEN 40
70       PRINT I
80       S = S + I
90 NEXT K
100 PRINT
110 PRINT S/N
120 DATA 0, 100
130 END
```

Fig. 6.26

For b) we introduce the variables K and S. K counts the random walks and S accumulates the total number of stops in all of them. This results in the program in Fig. 6.26. For c) we need to count the frequency of the duration I in L(I). Fig. 6.27 shows the program. It simulates N = 1000 random walks and determines the frequencies of duration of up to 99 steps. Larger durations are highly improbable.

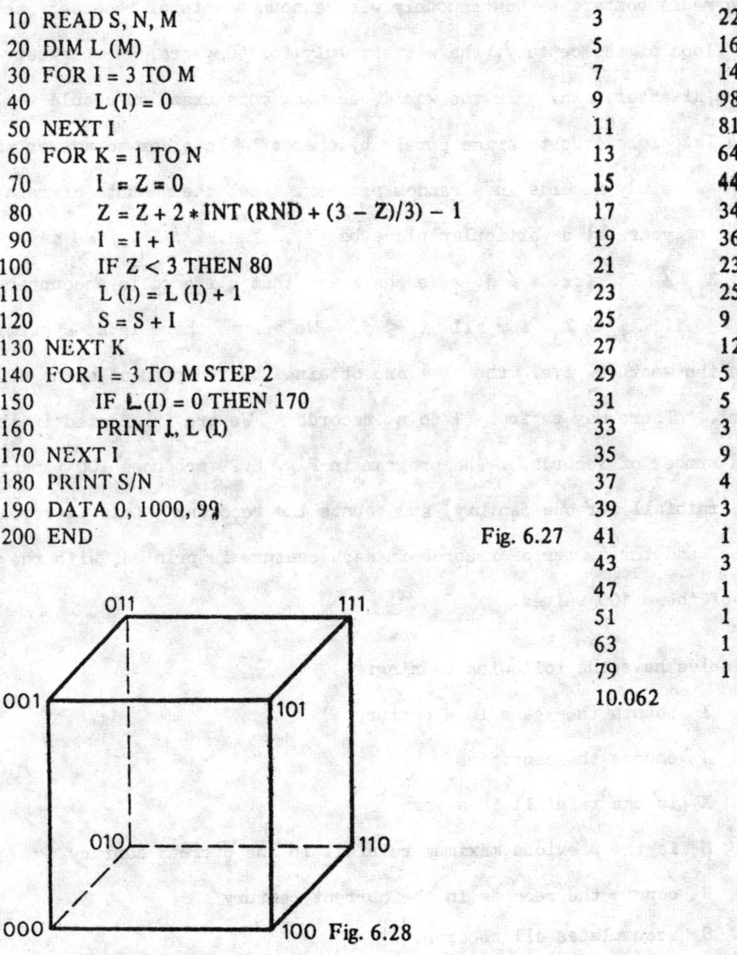

```
10 READ S, N, M
20 DIM L (M)
30 FOR I = 3 TO M
40     L (I) = 0
50 NEXT I
60 FOR K = 1 TO N
70     I = Z = 0
80     Z = Z + 2 * INT (RND + (3 - Z)/3) - 1
90     I = I + 1
100    IF Z < 3 THEN 80
110    L (I) = L (I) + 1
120    S = S + I
130 NEXT K
140 FOR I = 3 TO M STEP 2
150    IF L (I) = 0 THEN 170
160    PRINT I, L (I)
170 NEXT I
180 PRINT S/N
190 DATA 0, 1000, 99
200 END
```

Fig. 6.27

```
3     221
5     168
7     148
9     98
11    81
13    64
15    44
17    34
19    36
21    23
23    25
25    9
27    12
29    5
31    5
33    3
35    9
37    4
39    3
41    1
43    3
47    1
51    1
63    1
79    1
10.062
```


Fig. 6.28

Exercise:

6. We label the eight corners of a cube with triples composed of zeros and ones. (Fig. 6.28). The digits are stored in Y(1), Y(2), Y(3). The random walk on the cube can be simulated as follows: A random digit is chosen from $\{1, 2, 3\}$ and then Y(D) ← 1 - Y(D) is performed. Initially Y(1) = Y(2) = Y(3) = 0. We stop as soon as

$Y(1) + Y(2) + Y(3) = 3$. Write programs corresponding to Figs. 6.25 to 6.27.

9th Example: 'Records' in a random process

The news media bombard us continuously with announcements of records: the largest flood of the century, the wettest July for 50 years, the biggest aeroplane disaster, etc. Is the world becoming more crazy or should we expect a lot of records to arise purely by chance? In order to answer these questions we study records in a random process. Let the amounts of rain in the next n years at a particular place be $X_1, X_2 \ldots X_n$. We may suppose $X_i \neq X_j$ for $i \neq j$. We shall say that a record is encountered in year j if $X_i < X_j$ for all $i < j$. We assume there is no systematic trend in the weather, i.e. the X_i are obtained by independent spins of a 'spinner'. There may be from 1 to n records. We are interested in the expected number of records. The program in Fig. 6.29 produces 100 random numbers (rainfall for one century) and counts the records. This is repeated 100 times, and the number of records in each century is printed, with the average of these 100 values.

The variables have the following meaning:

- I counts the years in a century
- J counts the centuries
- X is the rainfall in a year
- M is the previous maximum rainfall in the current century
- R counts the records in the current century
- S accumulates all records.

The result is slightly over 5 records per century, i.e. records are relatively infrequent. But the news media observe a large number of random processes. It can be shown that the expected number of records in a century is 5·187 (see [5]).

```
10 S = 0
20 FOR J = 1 TO 100
30     M = R = 0
40     FOR I = 1 TO 100
50         X = RND
60         IF X < M THEN 90
70         R = R + 1
80         M = X
90     NEXT I
100    PRINT R;
110    S = S + R
120 NEXT J
130 PRINT
140 PRINT S/100
150 END
```

Fig. 6.29

```
4 7 3 8 8 6 5 7 4 4 6 8 6 5 4 4 5 2 6 5 6 5 8 4 8 3 2 3 6 5 4 4 5 2 5 5 2
10 7 10 2 2 9 5 4 5 7 6 6 5 3 8 5 3 8 3 8 4 4 4 7 2 8 3 5 3 5 5 4 2 8 2 4
6 8 9 5 3 6 6 10 6 5 4 3 4 6 6 3 8 7 6 3 5 7 4 3 2 3 7
5.17
```

<u>10th Example: Fishing in the Three shires lake</u>

Fig. 6.30 shows the Dreilandersee (three shires lake). The weights of the fishes in the lake are uniformly distributed in (0, 1), i.e. the function RND produces a fish. In order to conserve the fish, each county has a stop-rule for its anglers. Let G_1, G_2, G_3 . . . be the weights of the fishes which are caught by successive anglers. The stop-rules for Anchurien, Sikinien and Zentaurien are at the moment:

A. Stop as soon as $G_{n-1} < G_n$

S. Stop as soon as $G_1 + G_2 + . . . + G_n > 1$

Z. Stop as soon as $G_n > G_1$.

Let X be the number of fishes which an angler catches. We want to find the distribution of X and E(X) for rule A. For this we ascertain how often X takes the values 2, 3, 4, 5, . . . for 1000 anglers. If the average number in a catch is X, R(X) is increased by 1. S counts all the fish. V and N are the weights of the two successive fishes under consideration.

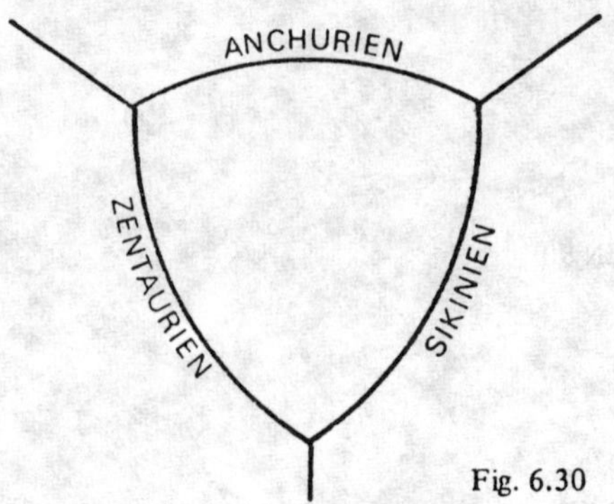

Fig. 6.30

```
 10 S = 0
 20 FOR X = 2 TO 10  R (X) = 0
 30 FOR I  = 1 TO 1000
 40       X = 2; V = RND; N = RND
 50       IF V < N THEN 70
 60       V = N; N = RND; X = X + 1; GO TO 50
 70       R (X) = R (X) + 1;  S = S + X
 80 NEXT I
 90 FOR X = 2 TO 10
100       IF R (X) < > 0 THEN PRINT X, R (X)/1000
110 NEXT X
120 PRINT
130 PRINT  "AVERAGE VALUE        ="S/1000
140 END
 2      0.478
 3      0.361
 4      0.120
 5      0.033
 6      0.008
    AVERAGE VALUE     .=2.732
```

Fig. 6.31

The program has an error which only appears when X becomes greater than 10. i.e. when a decreasing sequence $G_1 > G_2 > G_3 \ldots > G_{10}$ occurs.

The probability of such a decreasing sequence of length 10 is $q_{10} = \frac{1}{10!}$.

For 1000 anglers this will occur at least once with probability

$$1 - (1 - q_{10})^{1000} < 1000q_{10} = \frac{1}{3628 \cdot 8}$$

It is easily shown that $X = n$ with probability $p_n = \frac{n-1}{n!}$ and $E(X) = e = 2 \cdot 71828\ldots$ (see [5]). These values should be compared with the estimates obtained in Fig. 6.31.

The program for the stop-rule S is completely analogous. See Exercise 7. The reader should certainly try to solve this one. The result is surprising.

The simple stop-rule Z is particularly interesting. I wish to thank Herr A. Vogt of the Federal Statistical Office (E.S.A.) in Bern for drawing this to my attention. We deal with this rule first by calculation. Clearly $X > n$ precisely when $G_1 = \max(G_1, G_2 \ldots, G_n)$. The probability of this is

$$q_n = \frac{1}{n}, \quad n = 1, 2, 3 \ldots$$

Thus $X = n$ with probability $p_n = q_{n-1} - q_n$. i.e.

$$p_n = \frac{1}{n-1} - \frac{1}{n} = \frac{1}{n(n-1)}, \quad n = 2, 3, 4 \ldots$$

and $$E(X) = \sum_{n \geq 2} np_n = 1 + \frac{1}{2} + \frac{1}{3} + \ldots \text{ which is infinite.}$$

The infinite value is a surprise. The simulation presents some difficulties here, since with 1000 anglers it is to be expected that there is at least one who has caught over 1000 fish. This follows from $1000q_{1000} = 1$. (†)

Five simulations each involving 1000 anglers gave average catches of

17.126, 12.042, 6.297, 7.325, 5.548

Are these good estimates for infinity?

[(†) in more detail: Probability that a given angler catches under 1000 fish $= 1 - q_{1000} = 1 - \frac{1}{1000}$, so probability that 1000 anglers each catch under $1000 = (1 - \frac{1}{1000})^{1000} \simeq e^{-1} = 0{\cdot}3678\ldots$]

Exercises:

7. Write a program like that of Fig. 6.31 for the stop-rule S. Compare the two frequency tables. Conjecture?

8. We wish to simulate the stop-rule Z. For this we repeat the angling process 1000 times, determine the average numbers caught, and the relative frequencies of $X = 2, 3, 4, \ldots 10$, as well as $X > 10$. Write the corresponding program.

9. A particle starts at 0 in a symmetric random walk on the line and makes $2n = 1000$ steps. Write a program which prints the frequency of return to the origin. Compare with the expected value

$$E_{2n} = (2n + 1) \binom{2n}{n} 2^{-2n} - 1 = 24.2502$$

The proof of this formula may be found in [5], pp 50 and 225.

10. A particle starts at 0 on a line and performs a symmetrical random walk of 1000 steps. Write a program which prints the maximum distance from 0 which is reached.

11. A particle starts at 0 on a line and performs a symmetrical random walk. If it reaches 3 or −3, it stops. Let X be the number of steps to absorption. The random variable X can take values 3, 5, 7, 9, . . . We wish to estimate E(X) and the distribution of X.

    a) Write a program which prints the number of steps in each of 100 random walks, with their average.

    b) Write a program which carries out 1000 random walks, and prints the frequencies of the step-totals 3, 5, 7, 9, . . . with their averages.

(The exact values are $E(X) = 9$, and $p_n = \frac{1}{4} \left(\frac{3}{4}\right)^{\frac{n-3}{2}}$ for $n \in \{3,5,7,9,11,\ldots\}$)

12. A particle starts at O and performs a symmetrical random walk in 3D space. At each step it moves with probability $\frac{1}{6}$ to one of the six neighbouring points. The program in Fig. 6.32 simulates 100 random walks each of 10 steps in a clear and elegant way. Study the program until you understand it and then run it on the computer. Is it true that $E(D_n^2) = n$ for 3D space as well?

```
10 READ S, N, M
20 FOR I = 1 TO M
30       X (1) = X (2) = X (3) = 0
40       FOR K = 1 TO N
50             A = INT (3*RND) + 1
60             X (A) = X (A) + 2*INT (2*RND) – 1
70       NEXT K
80       D = X (1)↑2 + X (2)↑2 + X (3)↑2
90       S = S + D
100 NEXT I
110 PRINT S/M
120 DATA 0, 10, 100
130 END
```

Fig. 6.32

13. A particle starts at 0 and performs a random walk in space with steps of length 1, as follows: A sphere S of radius 1 is drawn about the current position (x, y, z). The particle moves from the centre of the sphere to a "randomly" chosen point ω of S. i.e. ω falls in a subset of S with probability

$$P(\omega \in M) = \frac{\text{area of M}}{\text{area of S}} = \frac{|M|}{4\pi}$$

Archimedes showed that a 'zone' of the Earth's surface ([between two parallels of latitude]) of height H has the area $2\pi h$ independent of the position of the zone. Because of this it is necessary to choose the height h of the point ω (above the equator) at random from (-1, 1) by h ← 2*RND - 1. The geographical longitude a of ω is randomly chosen between 0 and 2π by a ← 2π*RND. If we write r ← $\sqrt{1 - h^2}$ (fig. 6.33b), the substitutions x ← x + r cos a, y ← y + r sin a, z ← z + h represent a random step. Simulate 100 random walks each of 10 steps and estimate $E(D_n^2)$.

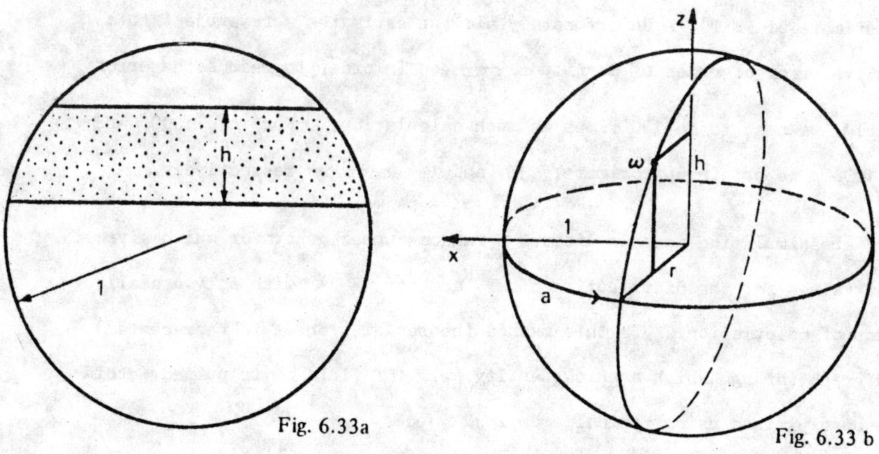

Fig. 6.33a

Fig. 6.33 b

6.3 Simulations without a random-number generator

Usually in a random process one is interested in some random variable T which takes the values 0, 1, 2, . . . with probabilities $p_0, p_1, p_2 \ldots$ The sequence p_n is called the distribution of T. The expected value of T is defined by

$$(1) \qquad E(T) = \sum_{i\geq 0} i.p_i = 0.p_0 + 1.p_1 + 2.p_2 + 3.p_3 + \ldots$$

We rearrange the right hand side of (1) as follows:

$$(2) \qquad E(T) = (p_1 + p_2 + p_3 + \ldots)+(p_2 + p_3 + p_4 + \ldots)+(p_3 + p_4 + p_5 + \ldots)+$$

If we write

$$(3) \qquad q_i = p_{i+1} + p_{i+2} + p_{i+3} + \ldots$$

we have

$$(4) \qquad E(T) = \sum_{i\geq 0} q_i = q_0 + q_1 + q_2 + q_3 + \ldots$$

Also q_i is the probability that $T > i$.

Earlier we have simulated random processes using a random number generator in which we have run through them n times. The relative frequency h_i of the outcome i was an estimate of p_i and the mean value $\sum ih_i$ was an estimate of E(T). Unfortunately all the estimates were subject to a relative error of order of magnitude $\frac{1}{\sqrt{n}}$. In order to reduce the error 10 fold, we need to do 100 times as much calculation. For n = 1000 and n = 10000 we obtain approximately 3% and 1% 'accuracy' respectively.

There are simulation methods without a random number generator which give exact values for the distribution of T and for E(T) with only a small amount of calculation. In this method the random process is represented by a graph through which a unit quantity [e.g. of fluid] is pumped. This will be explained using typical examples.

1st Example: Waiting time for a sequence of successes

A fair coin with sides 0 and 1 is thrown repeatedly. Let T be the waiting-time until the first appearance of the sequence 1111. We denote the probabilities that $T = n$ and that $T > n$ by p_n and q_n respectively. We seek the values of p_n and $E(T)$. The simulation process is represented by Fig. 6.34. We start in state 0 and move about the graph, respecting the transition probabilities, until we reach the state 1111, where the process stops. T counts the moves up to the stopping point.

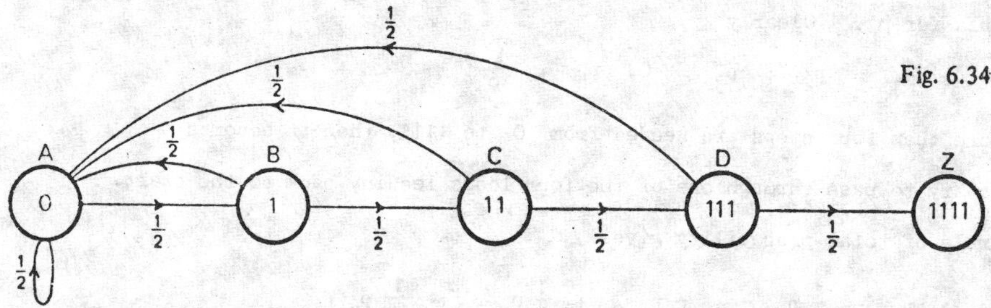

Fig. 6.34

We introduce the vector $\vec{M}_n = (a_n, b_n, c_n, d_n, z_n)$ in which $a_n, \ldots z_n$ are the probabilities that we find ourselves at time n (i.e. after the nth move) in the states 0, ... 1111 respectively. We interpret $\vec{M}_n$ as a subdivision of a unit quantity among the states of the graph. This material is pumped through the graph in discrete steps observing the transition-probability rules. Clearly $q_n = 1 - z_n$, and $p_n = \frac{d_{n-1}}{2}$. Fig. 6.35 pumps the material through the graph. Here (A, B, C, D, Z) are the current and (A1, B1, C1, D1, Z1) the next subdivisions. I counts the steps, and E calculates E(T) using (4). We break off as soon as the condition $Z < 1$ is no longer fulfilled.† The values p_4 to p_{20} printed by line 40 are omitted, since they appear in Fig. 6.36. It can be shown that $E(T) = 30$ (see [5] p 31). Our program gives slightly too large an estimate, through rounding error.

```
10 A = 1; B = C = D = Z = E = I = 0
20 A1 = (A + B + C + D)/2; B1 = A/2; C1 = B/2; D1 = C/2; Z1 = Z + D/2
30 E = E + 1 - Z; I = I + 1
40 IF ABS (I - 12) < = 8 THEN PRINT I, D/2
50 A = A1; B = B1; C = C1; D = D1; Z = Z1
60 IF Z < 1 THEN 20
70 PRINT "I=" I, "E=" E
80 END

I = 660      E = 30.00000001
```

Fig. 6.35

[† when the computer can no longer detect the difference]

Line 60 is particularly dangerous. By rounding errors so much 'material' can be lost, that $Z < 1$ remains true perpetually† and the computer gets caught in a loop. It is safer to replace $Z < 1$ by $1 - Z > 1E - 9$.

The above solution requires no previous knowledge. However, if one is familiar with probability theory one can derive a recursion formula for p_n. Clearly

$$(5) \qquad p_1 = p_2 = p_3 = 0, \quad p_4 = \frac{1}{16}$$

If more than four steps are needed from 0 to 1111, then it becomes necessary to pass through one of the four loops leading back to the start. The rule of total probability gives

$$(6) \qquad p_n = \frac{1}{2} p_{n-1} + \frac{1}{4} p_{n-2} + \frac{1}{8} p_{n-3} + \frac{1}{16} p_{n-4}$$

```
10 DIM P (400)
20 P (1) = P (2) = P (3) = 0   P (4) = 1/16; E = 1/4; I = 4 ;F = 1/16
30 I = I + 1; P (I) = P (I - 1)/2 + P (I - 2)/4 + P (I - 3)/8 + P (I - 4)/16
40 IF I < = 32 THEN PRINT I, P (I), P (I)/P (I   1)
50 E = E + I*P (I); F = F + P (I)
60 IF 1 - F > 1E - 6 THEN 30
70 PRINT "I=";I, "E=";E
80 END
```

Fig. 6.36

i	p_i	p_i/p_{i-1}
5	0.03125	0.5
6	0.03125	1.0
7	0.03125	1.0
8	0.03125	1.0
9	0.029296875	0.9375
10	0.0283203125	0.9666666667
15	0.02359008789	0.9638403990
20	0.01961612701	0.9637803392
25	0.01631191373	0.9637809778
30	0.01356427837	0.9637809881
31	0.01307299361	0.9637809877
32	0.01259950269	0.9637809877

I = 377 E = 29.99959731

The program in Fig. 6.36 is based upon (5) and (6). The variable E calculates $E(T)$ using (1). Initially $E = \frac{1}{4}$, since $4p_4 = \frac{1}{4}$, and in

† [even though the computer cannot detect very small differences].

row 50 the counter I begins at 5. The variable F sums the p_n. We break off as soon as F comes within 10^{-6} of 1. Next to p_n we also print $\frac{p_n}{p_{n-1}}$. Fig. 6.36 shows only an extract of the printed table. The quotient $\frac{p_i}{p_{i-1}}$ converges to the limit $\lambda = 0{\cdot}9637809077$ with convergence factor $\frac{1}{2}$. i.e. for large n we have $p_n \sim c\lambda^n$.

If we substitute $p_n = c\,\lambda^n$ in (6), we obtain

$$(7) \qquad \lambda^4 - \frac{\lambda^3}{2} - \frac{\lambda^2}{4} - \frac{\lambda}{8} - \frac{1}{16} = 0$$

Exercises:

1. Determine the positive solution of the equation (7).
2. A fair coin with sides 0 and 1 is tossed repeatedly. Let T be the waiting-time until the first appearance of the sequence 0011.
   (a) Verify that this random process may be represented by Fig. 6.37.
   (b) Write a program analogous to Fig. 6.35.
   (c) It can be shown that $p_n = p_{n-1} - \frac{p_{n-4}}{16}$ (see [5] p 80).
   Write a program analogous to Fig. 6.36.
   (d) Substitute $p_n = c\,\lambda^n$ in the recursion for p_n and find the largest positive solution of the equation for λ.

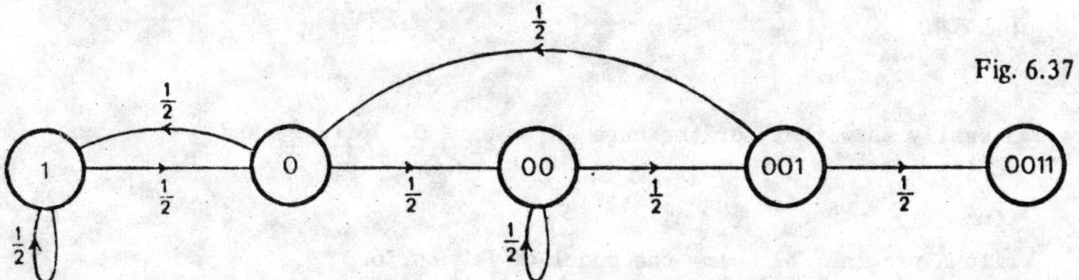

Fig. 6.37

2nd Example: Random walk on a cube.

We examine the symmetric random walk on a cube, which starts at 0 and finishes at 3 (Fig. 6.38) It is equivalent to the random walk on the graph in Fig. 6.39. We begin with a mass 1 at position 0, and continue circulating it until the whole mass is at 3. Let the masses concentrated at 0, 1, 2, 3, be A, B, C, D and one step later let them be A1, B1, C1, D1. Initially A = 1, B = C = D = 0. Let T be the time taken to reach 3 from 0. The program in Fig. 6.40 gives the first 25 terms of the distribution of T and also E(T). To save space the <u>distribution</u> is omitted. See, however, Exercise 3 following.

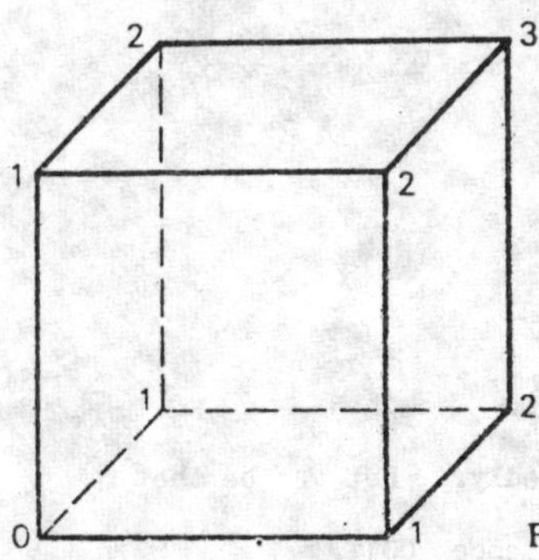

Fig. 6.38

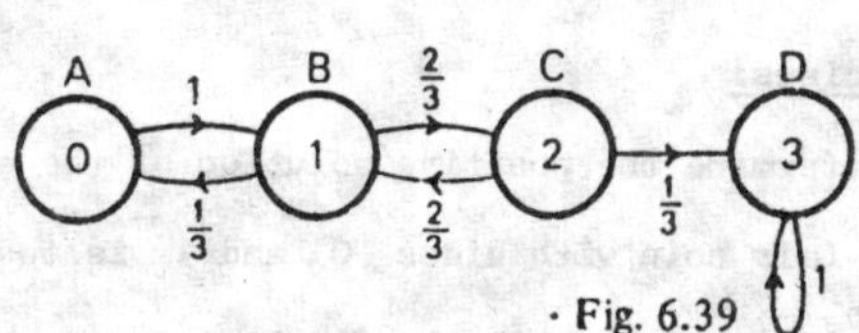

Fig. 6.39

```
10 A = 1; B = C = D = I = 0 ; E = 0
20 I = I + 1; A1 = B/3; B1 = A + 2*C/3; C1 = 2*B/3; D1 = D + C/3
30 IF I <=25 THEN PRINT I, C/3
40 E = E + 1 - D
50 A = A1; B = B1; C = C1; D = D1
60 IF D < 1 THEN 20
70 PRINT "I=";I, "E=";E
80 END
I = 193          E = 10
```

Fig. 6.40

<u>Exercise</u>:

3. One can easily show that for the cube $p_1 = p_2 = 0$, $p_3 = \frac{2}{9}$ and

$$p_n = \frac{7}{9} p_{n-2} \quad \text{for } n > 3.$$

(a) Write a program following the model of Fig. 6.36.

(b) Find a closed formula for p_n.

(c) Write a program to calculate E(T) using the formula in (b) and (1) above.

(d) Determine the smallest n for which $q_n < 10^{-3}$.

<u>Reminder</u>: $q_n = p_{n+1} + p_{n+2} + \ldots = 1 - p_1 - p_2 - \ldots - p_n$

3rd Example: A game with a coin

Abel and Cain take part in the following game. A fair coin is thrown until either the sequence 111 or the sequence 101 is obtained. In the first case Cain wins, in the second case Abel wins. We require the probability of a win for each player, and the average length of a game.

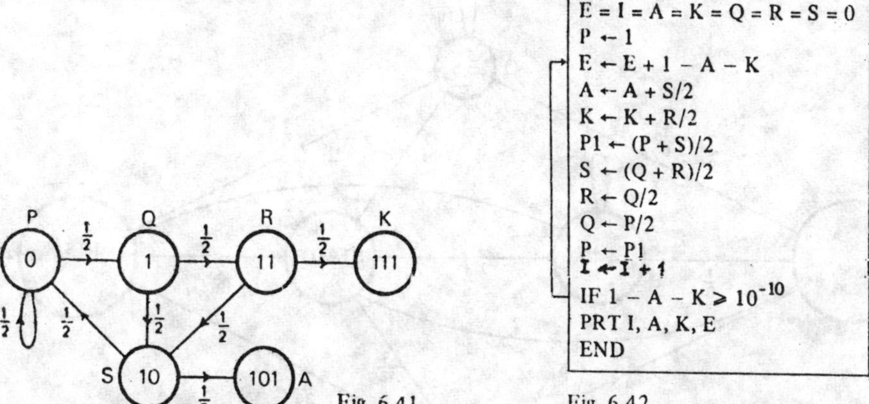

Fig. 6.41

Fig. 6.42

The game can be represented by the graph in Fig. 6.41. Each possible play of the game is a random walk on this graph, which begins in state 0 and ends in one of the states 111 or 101. Initially the mass 1 is concentrated in state 0. Let the amount in states 101 and 111 after n steps be A_n and K_n respectively. Then A_n and K_n are the probabilities that Abel and Cain respectively win the game in n steps. The probability that the game lasts longer than n steps is $1 - A_n - K_n$. In Fig. 6.41 and 6.42, P, Q, R, S, K and A denote the current masses in the states 0, 1, 11, 10, 111, 101. I counts the steps and E accumulates the average game length. The computer gives

$$I = 114, \quad A = 0{\cdot}6, \quad K = 0{\cdot}4, \quad E = 6{\cdot}8$$

4th Example: We repeat the rules of the game of crap.

1. Roll two dice and determine the sum of the scores S. S = 7 or S = 11 wins at once. S = 2 or S = 3 or S = 12 loses at once.
2. For every other sum, this sum is called 'point' P and the game continues until either S = 7 (lose) or S = P (win) occurs.

The following table shows the probabilities of the sums 2 to 12:

S	2	3	4	5	6	7	8	9	10	11	12
P (S)	$\frac{1}{36}$	$\frac{2}{36}$	$\frac{3}{36}$	$\frac{4}{36}$	$\frac{5}{36}$	$\frac{6}{36}$	$\frac{5}{36}$	$\frac{4}{36}$	$\frac{3}{36}$	$\frac{2}{36}$	$\frac{1}{36}$

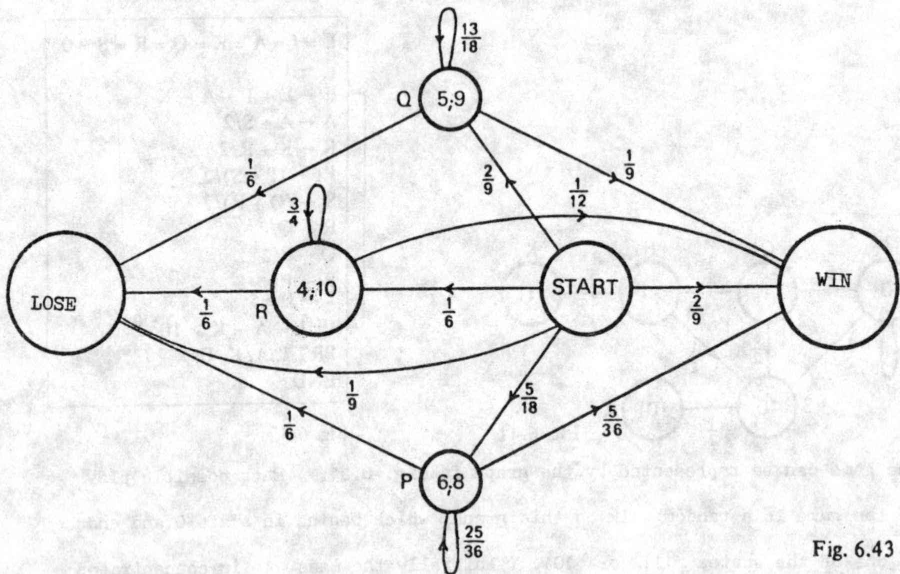

Fig. 6.43

The game can be represented by the graph in Fig. 6.43. We require the probability of winning G and the expected length of a game E. The meanings of the variables P, Q, R, V, G are obvious from Fig. 6.43. This time we will start with a mass of 18. After one step the variables P,Q,R,V,G,E have the values 5, 4, 3, 2, 4, 18. Finally we divide G and E by 18. The result is:

$$G = 0{\cdot}4929292929 = \frac{244}{495} \qquad E = 3{\cdot}375757575 = 3\ \frac{62}{165}$$

These values are exact. (see [5]).

```
10 READ P, Q, R, V, E
20 E = E + P + Q + R
30 V = V + (P + Q + R)/6
40 P = 25*P/36
50 Q = 13*Q/18
60 R = 3*R/4
70 IF P + Q + R >= 1E - 10 THEN 20
80 PRINT 1 - V/18, E/18
90 DATA 5, 4, 3, 2, 18
100 END
.4929292929    3.375757576
```

Fig. 6.43

Exercises:

4. Let T be the number of steps from 1 to 4 in Fig. 6.44. Write a program analogous to Fig. 6.35 which determines E(T).

5. Let T be the number of steps from 0 to 5 in Fig. 6.45. Write a program to determine E(T).

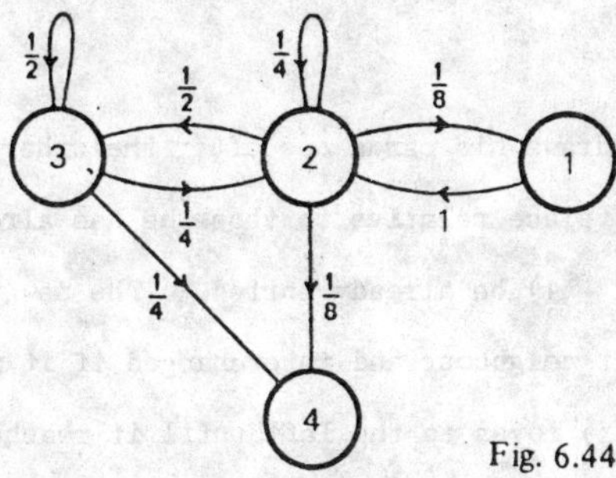

Fig. 6.44

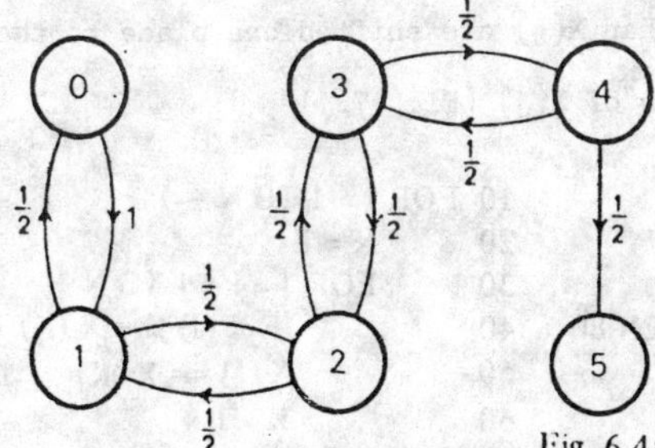

Fig. 6.45

7. SORTING

Let n real numbers be stored in X(1) to X(n). The numbers are to be arranged in increasing order, so that $X(i) \leqslant X(i + 1)$ for $i = 1, \ldots, n-1$. This rearrangement is called <u>sorting</u>. At first sight sorting appears trivial and without interest. In fact the exact opposite is true. It is estimated that about one quarter of the running time of all computers is concerned with sorting. This shows the great economic importance of sorting. On the other hand sorting algorithms are among the most interesting, ingenious and difficult of all algorithms. Hundreds of sorting algorithms have been constructed. With the slowest the sorting time is proportional to n^2, with the fastest, to $n. \log_2 n$. We consider slow algorithms because they are simple. We obtain data for sorting by the instruction

```
FOR I = I TO N     X(I) = INT (N*RND)     NEXT I
```

This line stores random numbers from $\{0, 1, \ldots, n - 1\}$ in X(1) to X(n).

a) <u>Sorting by insertion</u>

We proceed like a bridge-player who draws his cards one after the other and inserts each new card in its correct place relative to those he has already drawn. Let the numbers X(1) to X(j - 1) be already sorted. The new number X(j) is compared with its left neighbour and interchanged if it is the smaller of the two. In this way X(j) moves to the left until it reaches the position i + 1 for which $X(i) \leqslant X(j)$. Thus those of the numbers from X(1) to X(j - 1) which are greater than X(j) are shifted one place to the right to allow room for the insertion of X(j) (Fig. 7.1)

```
10 FOR J = 2 TO N
20     I = J - 1
30     X = X (J)
40     IF X > = X (I) THEN 80
50     X (I + 1) = X (I)
60     I = I - 1
70     IF I > 0 THEN 40
80     X (I + 1) = X
90 NEXT J
100 END
```

Fig. 7.1

```
10 FOR I = 1 TO N - 1
20     K = I
30     FOR J = I + 1 TO N
40         IF X (J) > = X (K) THEN 70
50         X (J) == X (K)
60         K = J
70     NEXT J
80 NEXT I
90 END
```

Fig. 7.2

b) <u>Sorting by selection</u>

First set $i = 1$. Then the smallest element is chosen from among $X(i)$ to $X(n)$, and interchanged with $X(i)$. The ith element is now in its correct place. Now $i \leftarrow i + 1$ is performed and the step is repeated as long as $i < n$ (Fig. 7.2).

c) <u>Sorting by interchange</u>

The sorting algorithm in Fig. 7.3 is called a <u>bubble-sort</u>. It is the easiest to understand and also the least efficient sorting algorithm.

```
10 FOR J = 1 TO N - 1
20      FOR I = 1 TO N - J
30          IF X (I) < = X (I + 1) THEN 50
40          X (I) == X (I + 1)
50      NEXT I
60 NEXT J
70 END
```

Fig. 7.3

If the algorithm is used with a list which is already sorted, it still requires $\frac{n(n - 1)}{2}$ comparisons. The algorithm in Fig. 7.4 avoids this. 100 random numbers are placed in X(1) to X(100). The auxiliary variable F is called a 'flag'. At the beginning of each pass, F is set equal to 1. If two neighbours are interchanged during a pass, F is set to 0 and line 90 begins a new pass. If the list is already sorted, the computer detects this in line 90 and prints the sorted list.

```
 10 DIM X (100)
 20 FOR I = 1 TO 100  X (I) = INT (100 * RND)
 30 F = 1
 40 FOR I = 1 TO 99
 50      IF X (I) < = X (I + 1) THEN 80
 60      X (I) == X (I + 1)
 70      F = 0
 80 NEXT I
 90 IF F = 0 THEN 30
100 FOR I = 1 TO 100 PRINT X (I);
110 END
```

```
1 3 5 5 5 7 8 12 14 14 16 18 18 18 19 20 21 21 22 22 24 24 26 27 28 28 28
29 30 30 31 31 32 33 34 34 34 36 38 38 38 40 41 42 43 44 47 47 47 48 48
49 49 50 51 51 53 53 55 55 59 59 59 59 60 60 62 62 62 63 65 65 67 69 72
72 73 76 76 77 77 77 77 78 78 80 82 83 84 86 87 93 93 95 95 95 96 98 98 99
```

Fig. 7.4.

Exercises:

1. Sort the list 6, 3, 7, 4, 8, 5, 2, 1 by hand using each of the algorithms 7.1 to 7.3.
2. Use each of the algorithms 7.1 to 7.3 to sort 100 random numbers, and compare the computing time.

   Do not forget the instruction DIM X(100).

d) Sorting by frequency-count

It often happens that the numbers to be sorted lie in a small interval $u \leq X(i) \leq v$, for example, $1 \leq X(i) \leq 10$. In one pass through the list we can count how often each of 1, 2, . . . 10 occurs, and then, in principle, the list is sorted. Fig. 7.5 shows a possible program. We store the frequencies of the numbers u, u + 1, . . . , v in the array Z(u), Z(u + 1), . . . Z(v). Finally the sorted X-array is stored in the S-array S(1) to S(n). This is a very quick program.

Commentary

10. reserves storage for the Z-list (frequency list) and initialises it.

20. counts the frequency of the numbers u to v in the X-list.

30 determines the cumulative frequencies - i.e. Z(i) gives the number of elements of the X-list which are $\leq$ i.

40-60 produce the S-list which is a sorted X-list.

```
10 FOR I = U TO V         Z (I) = 0          NEXT I
20 FOR J = 1 TO N         Z (X (J)) = Z (X (J)) + 1     NEXT J
30 FOR I = U + 1 TO V   Z (I) = Z (I) + Z (I - 1)   NEXT I
40 FOR J = N TO 1 STEP - 1
50        I = Z (X (J)); S (I) = X (J); Z (X (J)) = I - 1
60 NEXT J
70 END
```

Fig. 7.5

Exercises:

3. Write a program which stores the throws of a die in X(1) to X(100), sorts the X-list as in Fig. 7.5 and prints the sorted S-list.

4. Write a program which produces 100 throws of 2 dice and stores the sum of the scores on the ith throw in X(i). Next the X-list is to be sorted and the sorted S-list is to be printed.

```
 10 DIM X (100), Y (100)
 20 Z = (SQR (5) – 1)/2
 30 FOR I = 1 TO 100     X (I) = I
 40 FOR I = 1 TO 100     Y (I) = I * Z – INT (I * Z)
 50 FOR I = 2 TO 100
 60       FOR J = 1 TO I – 1
 70             IF Y (J) < = Y (I) THEN 90
 80             Y (I) == Y (J); X (I) == X (J)
 90       NEXT J
100 NEXT I
110 FOR I = 1 TO 100 PRINT X (I);
120 END
```

Fig. 7.6

89 34 68 13 47 81 26 60 5 94 39 73 18 52 86 31 65 10 99 44 78 23 57 2 91 36
70 15 49 83 28 62 7 96 41 75 20 54 88 33 67 12 46 80 25 59 4 93 38 72 17 51
85 30 64 9 98 43 77 22 56 1 90 35 69 14 48 82 27 61 6 95 40 74 19 53 87 32
66 11 100 45 79 24 58 3 92 37 71 16 50 84 29 63 8 97 42 76 21 55

e) The golden permutation

We consider an interesting application of sorting. Let $z = \frac{\sqrt{5} - 1}{2}$ (ratio of the golden section). The n pairs of numbers $(i, iz - [iz])$, $i = 1, 2, \ldots n$ are sorted so that their second coordinates are increasing. Then the first coordinates are printed. There results a so-called "golden permutation" of the numbers 1 to n. It has amazing properties and can be taken as a random permutation. Fig. 7.6 prints the golden permutation of the numbers 1 – 100.

We enumerate some of the surprising properties of this permutation.

a) Only three possible differences occur between neighbouring elements of the permutation - these are 34, 55 and 89.

b) If we select two elements whose difference is 1, they are separated by 31, 38, or 61 other elements.

c) Suppose a random sample of 10 elements is to be chosen from a sequence of consecutive integers, for example 38, 39, . . ., 77. Then we start anywhere in the table, for example at 78, and choose successively those elements which lie in the interval 38 to 77. We obtain 57, 70, 49, 62, 41, 75, 54, 67, 46, 59. This sample is unusually uniformly distributed. Between neighbours only the differences 13, 21, 34 occur. Consider the interval $[38, 77]$ on the number-line. If the points 57, 70, 49, . . . are plotted successively in this interval, the latest point falls in one of the currently largest free intervals.

f) Sorting by merging (J. von Neumann 1945)

Two previously-sorted sequences $x_1 \le x_2 \le \ldots \le x_n$ and $y_1 \le y_2 \le \ldots \le y_n$ can be merged into order to give a sequence $z_1 \le z_2 \le \ldots \le z_{2n}$ using 2n - 1 comparisons. The reader is invited to study this obvious procedure for himself. Suppose a sequence of 16 numbers is to be sorted. First 8 sorted pairs are produced. These are merged in pairs to give four sorted sequences of 4. Merging in pairs again gives 2 sorted sequences of 8, and a further merging gives one sorted sequence of 16. The numerical example of Fig. 7.7 makes the procedure clear. The total number of comparisons is

$$V_{16} = 8 + 4 \cdot 3 + 2 \cdot 7 + 1 \cdot 15 = 49.$$

7	11	12	6	2	8	3	1	13	9	16	15	5	14	4	10
7	11	6	12	2	8	1	3	9	13	15	16	5	14	4	10
6	7	11	12	1	2	3	8	9	13	15	16	4	5	10	14
1	2	3	6	7	8	11	12	4	5	9	10	13	14	15	16
1	2	3	4	5	6	7	8	9	10	11	12	13	14	15	16

Fig. 7.7

Now let a sequence of $n = 2^m$ elements be sorted. By the argument used for the case n = 16, the number of comparisons required is

$$V_n = \frac{n}{2} \cdot 1 + \frac{n}{4} \cdot 3 + \frac{n}{8} \cdot 7 + \ldots + \frac{n}{2^m}(2^m - 1) = n \sum_{i=1}^{m} \left(1 - \frac{1}{2^i}\right) = m.n - n\left(1 - \frac{1}{2^m}\right)$$

(1) $$V_n = n \log_2 n - n + 1$$

An analogous argument shows that the total time required for sorting is proportional to $n.\log_2 n$.

If n is not a power of 2, it can be shown that the total number of comparisons is

$$(2) \qquad V_n = n \lceil \log_2 n \rceil - 2^{\lceil \log_2 n \rceil} + 1$$

Here $\lceil x \rceil$ is the smallest whole number $\geq x$. If n is a power of 2, then (2) is transformed into (1), since $\log_2 n$ is a whole number.

We forgo a BASIC program, since it is very long.

<u>Exercises:</u>

5. Calculate V_{20} using (2). Try to sort a sequence of 20 numbers using V_{20} comparisons.

6. In practice sequences with as many as 10^5 or 10^6 members must be sorted. Slow algorithms like a) - c) are useless in such cases. Faster algorithms like f) are required.
   How many comparisons are required by the algorithms c) and f) respectively for $n = 10^5$?

g) <u>Shellsort. Sorting by diminishing increment</u>

Insertion sort is slow because it exchanges only adjacent elements. Shellsort is a simple extension of insertion sort which allows exchanges of elements that are far apart. This speeds up sorting considerably.

Replace in insertion sort every occurrence of "1" by "H" and "2" by "H+1". You will get a program which rearranges a file so that every Hth element yields a sorted file. Such a file is called H-sorted. By H-sorting an array for any sequence of H-values ending in 1 you will get a sorted array. This is <u>Shellsort</u>. We will use the sequence 1, 4, 13, 40, ... 3 H+1, ... which has been shown empirically to do uniformly well.

```
10   INPUT N,M : DIM X(N): H = 1
20   FOR I = 1 TO N: X(I) = INT(M*RND(1)): NEXT I
30   H = 3*H+1 : IF H < = N  THEN 30
40   H = INT(H/3)
50   FOR J = H + 1  TO  N
60   V = X(J) : I = J
70   IF  X(I-H) > V  THEN  X(I) = X(I-H) : I = I-H : IF  I>H  THEN  70
80   X(I) = V
90   NEXT J
100  IF  H>1  THEN  40
110  FOR  I = 1  TO  N : PRINT X(I) ; " " ; : NEXT I
120  END
```

Fig. 7.8

Comment: Line 20 generates N random digits from 0,1,...,M-1 and stores them in the array X(1) to X(N). Line 30 steps through the H-sequence 1, 4, 13, 40, ... until H N is satisfied. Line 40 makes H = N again. Lines 50 to 90 H-sort the file. For H = 1 this is straight Insertion sort. Line 110 prints the sorted file.

Problems:

7. Generate a) 100 b) 200 c) 500 d) 1000 random integers from 0 to 99, sort them with Insertion sort as well as Shellsort and compare the running times.

8. THE EIGHT QUEENS PROBLEM

To conclude we consider the following famous and instructive problem.

> Place 8 queens on an 8 x 8 chessboard so that no one can take any of the others, i.e. so that no two queens are in the same row, column or diagonal.

The problem was first proposed by Max Bezzel in 1848 in a chess magazine, and was at first unnoticed. Then it was published by Dr. Nauck in the "Illustrierten Zeitung" (Illustrated Newspaper) and aroused great interest. Gauss noticed the problem in this newspaper and spent a lot of time on it. On 21.9.1850 Dr. Nauck published all solutions received. Gauss had found only 72 solutions. We will construct an algorithm which provides all the solutions.

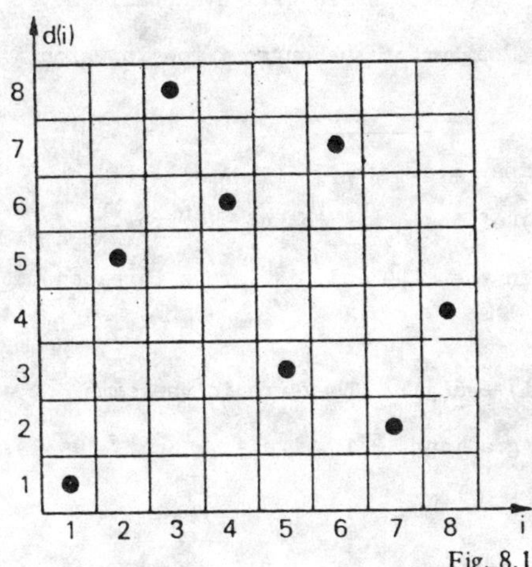

Fig. 8.1

Column i	1	2	3	4	5	6	7	8
Row d(i)	1	5	8	6	3	7	2	4

Fig. 8.2

The solution in Fig. 8.1 can be given as in Table 8.2, or even more briefly as the permutation

1 5 8 6 3 7 2 4

The computer will print all solutions in this form.

A queen must be placed in each row and in each column. We place the queens column by column, beginning with the first, and set d(1) = 1. Then we slide the second queen up the second column, from bottom to top until we find the first square which is not threatened. The second queen is placed on this square. Then we slide the third queen up the third column from bottom to top until we find a square for it which is not threatened by the other two queens, and so on.

If we cannot place a queen, because every square in the current column is threatened, we leave it aside and try to move the previous queen forward to a square which is not threatened by earlier queens. As soon as we find a solution, it is recorded, the last queen is removed from the board, and the previous queen is moved forward. A little reflection shows that we shall eventually find all solutions. Even the first queen will eventually move to the eighth square. Let i be the number of the current row in which a queen is to be placed.

Let j run through the previous columns i.e. j = 1 to i - 1.
Let d(i) be the ordinate (row-number) of the queen in the ith row.
How does one test whether the queens in the ith and j^{th} rows threaten each other?
The queens are in the same row if d(i) = d(j). They are in the same diagonal if the line joining them has gradient ± 1 i.e.

$$\frac{d(i) - d(j)}{i - j} = \pm 1$$

Since $i > j$ this can be written

$$|d(i) - d(j)| = i - j$$

From this we obtain the amazingly short program in Fig. 8.3, which we will explain in detail.

The lines 10 to 60 place the eight queens.
Line 20: the ith queen begins at the bottom of the column.
Lines 30 to 50 test whether the ith queen is threatened by earlier queens.
If so, we go to 90 and push the queen up one square. Line 100 tests whether

the queen is still on the board after this move. If so, we go back to 30 and test whether it is on a safe square. If not, lines 110 and 120 determine whether there exists an earlier queen. If so $(i \neq 0)$ we go to 90 and move it forward. If not, we have finished $(i = 0)$. As soon as $i = 8$ in 60, we have a solution. In 70 the solution is printed as an 8 place number. In 80 we go to the previous queen (column 7) and begin to move it upwards.

Fig. 8.3 shows that there are 92 different solutions.

We will regard two solutions as equivalent when they can be obtained from each other by rotation or reflection. A square can be superimposed on itself in eight ways. So in general there are seven other solutions equivalent to any given solution.

For example the following eight solutions are equivalent:

15863724	82417536	57263148	36428571
42736851	63571428	84136275	17582463

A solution in the first row is transformed into the next by a clockwise rotation of 90°. The solutions in the second row are obtained from those above by reflection in a vertical axis.

```
 10 FOR I = 1 TO 8
 20      D (I) = 1
 30      FOR J = 1 TO I - 1
 40           IF D (I) = D (J) OR ABS (D (I) - D (J) ) = I - J THEN 90
 50      NEXT J
 60 NEXT I
 70 PRINT D (1)*10↑7 + D (2)*10↑6 + D (3)*10↑5 + D (4)*10↑4 + D (5)*10↑3
                                  + D (6)*10↑2 + D (7)*10 + D (8)
 80 I = I - 1
 90 D (I) = D (I) + 1
100 IF D (I) <= 8 THEN 30
110 I = I - 1
120 IF I <> 0 THEN 90
130 END
```

15863724 16837425 17468253 17582463 24683175
25713864 25741863 26174835 26831475 27368514
27581463 28613574 31758246 35281746 35286471
35714286 35841726 36258174 36271485 36275184
36418572 36428571 36814752 36815724 36824175
37285146 37286415 38471625 41582736 41586372
42586137 42736815 42736851 42751863 42857136
42861357 46152837 46827135 46831752 47185263
47382516 47526138 47531682 48136275 48157263
48531726 51468273 51842736 51863724 52468317
52473861 52617483 52814736 53168247 53172864
53847162 57138642 57142863 57248136 57263148
57263184 57413862 58413627 58417263 61528374
62713584 62714853 63175824 63184275 63185247
63571428 63581427 63724815 63728514 63741825
64158273 64285713 64713528 64718253 68241753
71386425 72418536 72631485 73168524 73825164
74258136 74286135 75316824 82417536 82531746
83162574 84136275

Fig. 8.3

Altogether there are twelve non-equivalent solutions. We choose from each equivalent class the "minimum" representative and obtain

15863724 16837425 24683175 25713864

25741863 26174835 26831475 27368514

27581463 35281746 35841726 36258174

Now 12 x 8 = 96, not 92. The cause of this is the symmetrical solution no. 10 (Fig. 8.4) which is transformed into itself by a half-turn. Its equivalence class contains only four solutions:

35281746 64718257 64718253 75281746

The program in Fig. 8.3 is wasteful. The computing time can be substantially reduced by placing the queen in the first column only in positions 1, 2 and 3. The remaining solutions are then obtained by rotations and reflections.

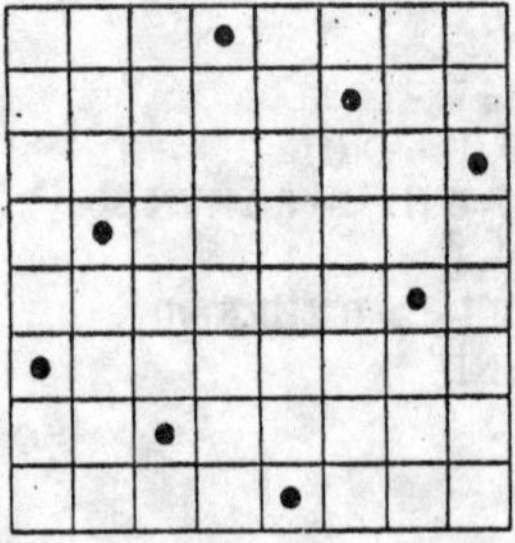

Fig. 8.4

9. SOLUTIONS OF SELECTED EXERCISES

1.1

1. $f(A) = A^{10}$
2. 1^2 , 2^2 , 3^3 , . . . , 10^2 .
3. See Fig. 1. Z shows how often the result is printed. The series obtained is $6n \pm 1$. The proof is not difficult.
4. Fig. 2 shows one of many possible solutions.
5. T ← A; A ← B; B ← C; C ← D; D ← E; E ← T.
6. A and B are interchanged.
7. 1^3 , 2^3 , 3^3 , . . . , 11^3.
8. See Fig. 2a.
9. $Y = f(X) = [\sqrt{X}]$.

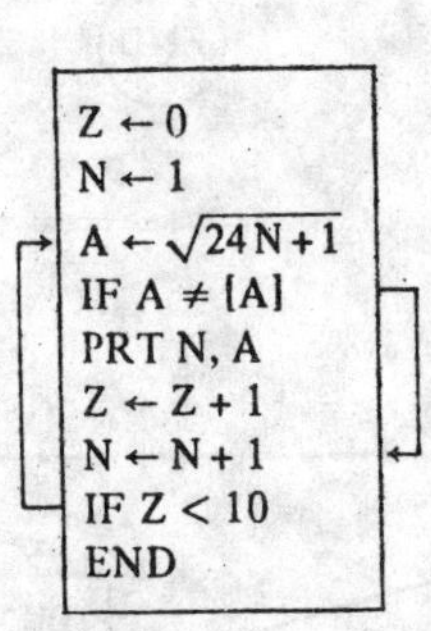

Fig. 1

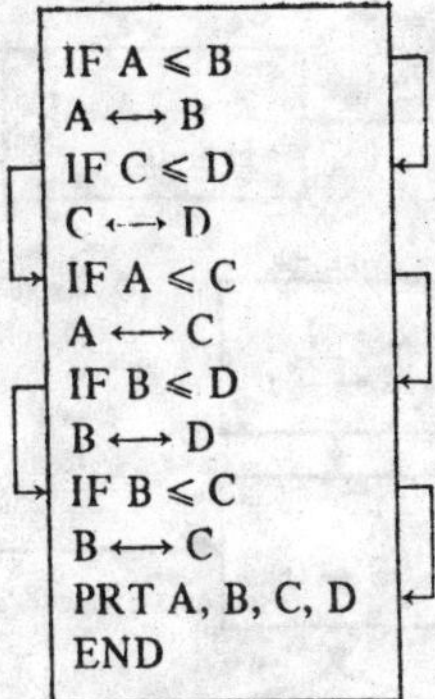

Fig. 2

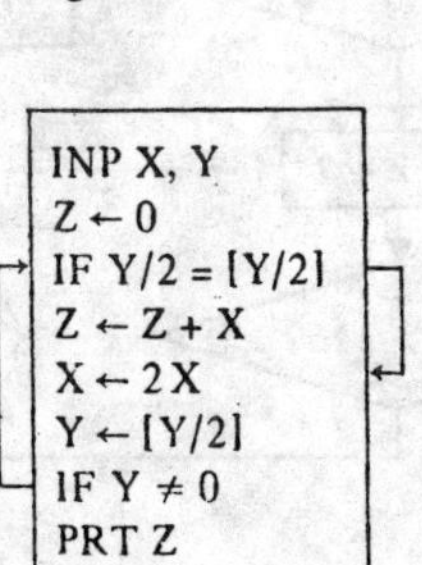

INP X, Y
Z ← 0
IF Y/2 = [Y/2]
Z ← Z + X
X ← 2X
Y ← [Y/2]
IF Y ≠ 0
PRT Z
END

Fig. 2a

1.2

2. See Fig. 3.

3. See Fig. 4. M is the current record number of steps.

4. See Fig. 5a and 5b.

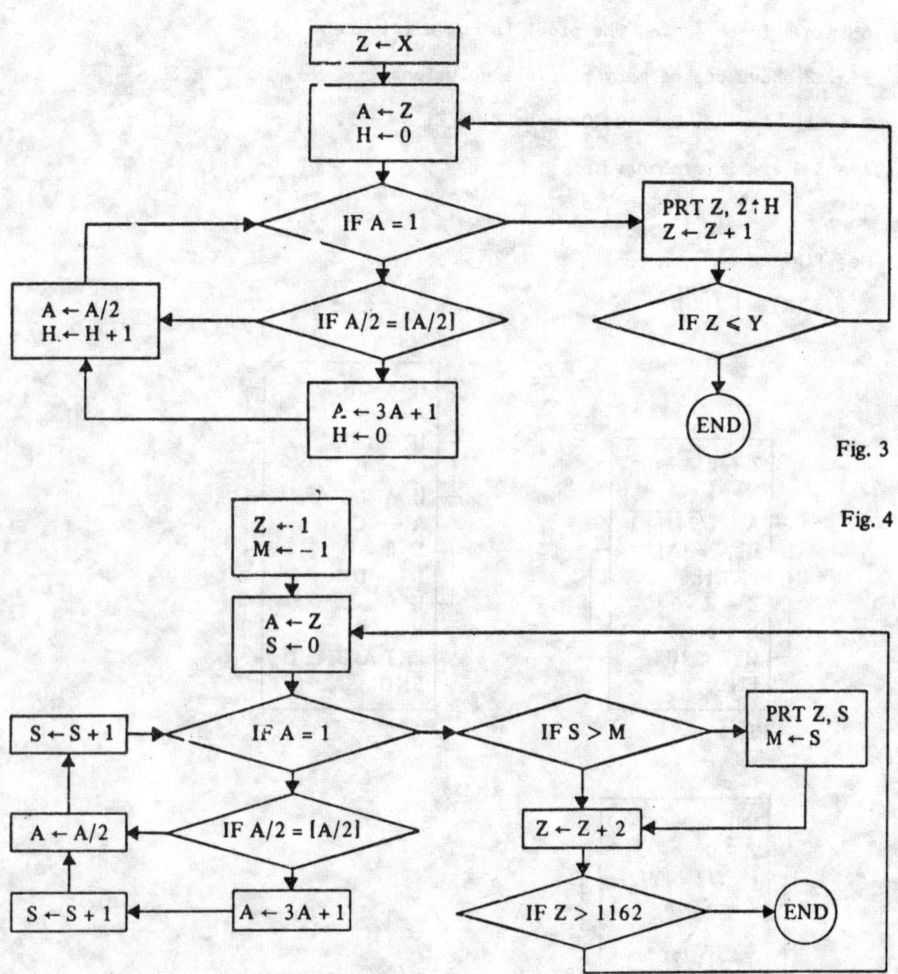

Fig. 3

Fig. 4

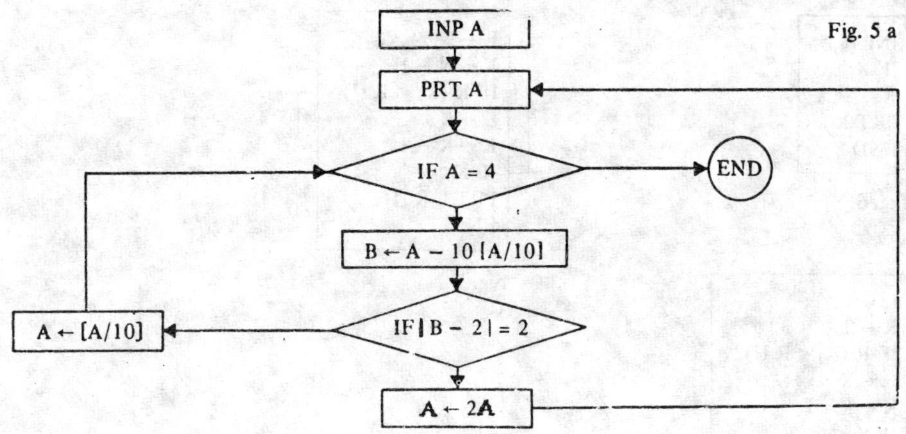

Fig. 5 a

```
 10 INP A
 20 PRINT A;
 30 IF A = 4 THEN 100
 40 B = A – 10*INT (A/10)
 50 IF ABS (B – 2) = 2 THEN 80
 60 A = 2*A
 70 GO TO 20
 80 A = INT (A/10)
 90 GO TO 20
100 END
```

Fig. 5 b

1.3.1

2. $m \leftarrow (x + y + |x - y|)/2$; $m \leftarrow (m + z + |m - z|)/2$; $m \leftarrow (m + u + |m - u|)/2$. The first two instructions give max (x,y,z) and all three give max (x,y,z,u).

3. In Fig. 1.24 and 1.25 $\geqslant$ must be replaced by $\leqslant$. In Fig. 1.26 $\leqslant$ must be replaced by $\geqslant$ in the third row.

4. See Fig. 6.

5. a) See Fig. 7 b) See Fig. 8

6. See Fig. 9

7. See Fig. 10

8. See Fig. 11. The minimum U and the maximum V are printed.

```
INP X
IF X ≥ 0
X ← – X
PRT X
END
```

Fig. 6

```
I ← N
K ← N – 1
IF R (K) ≤ R (I)
I ← K
K ← K – 1
IF K ≥ 1
PRT I, R (I)
END
```

Fig. 7

```
I ← 1
K ← 2
IF R (K) ≤ R (I)
I ← K
K ← K + 1
IF K ≤ N
M ← R (I)
IF R (I) ≠ M
PRT I, M
I ← I + 1
IF I ≤ N
END
```

Fig. 8

```
10 I = 1
20 FOR K = 2 TO N
30        IF R (K) < = R (I) THEN 50
40        I = K
50 NEXT K
60 PRINT I, R (I)
70 END
```

Fig. 9

```
10 I = J = 1
20 FOR K = 2 TO N
30        IF R (K) < = R (I) THEN 50
40        I = K
50        IF R (K) > = R (J) THEN 70
60        J = K
70 NEXT K
80 PRINT I; R (I), J; R (J)
90 END
```

Fig. 10

```
10 U = 1
20 V = 0
30 FOR I = 1 TO 100
40        R = I*SQR (2) – INT (I*SQR (2))
50        IF R < = V THEN 80
60        V = R
70        GO TO 100
80        IF R > = U THEN 100
90        U = R
100 NEXT I
110 PRINT U, V
120 END
```

Fig. 11

1.3.2

2. See Fig. 12

3. See Fig. 13

4. $1+1.2+1.2.3+ \dots +1.2. \dots .n = 1 + 2(1 + 3(1 + \dots n(1) \dots))$.

Fig. 14 evaluates this bracketed expression. Run this program for $n = 6$.

5. See Fig. 15.

6. See Fig. 16. From n = 8 the rounding errors become larger. Actually the sequence is monotone increasing with limit $e^{-1} = 0.3678794412$.

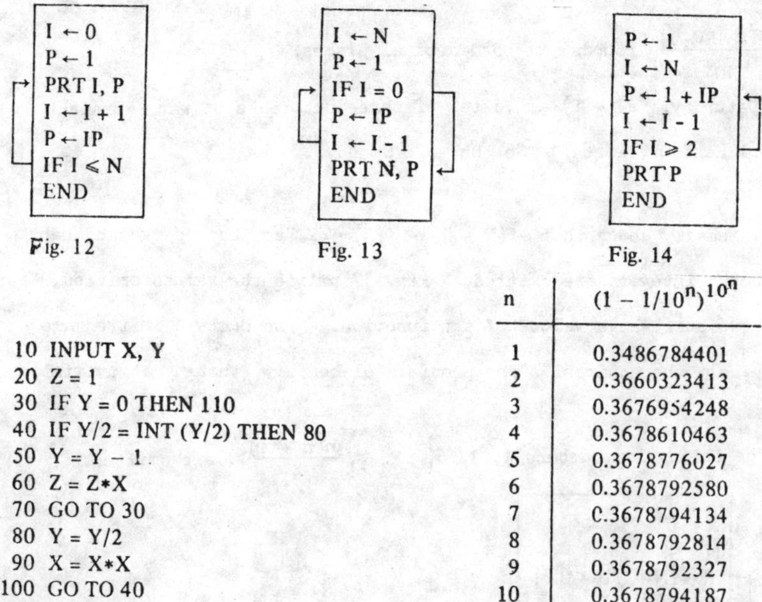

Fig. 12

Fig. 13

Fig. 14

```
10 INPUT X, Y
20 Z = 1
30 IF Y = 0 THEN 110
40 IF Y/2 = INT (Y/2) THEN 80
50 Y = Y - 1
60 Z = Z*X
70 GO TO 30
80 Y = Y/2
90 X = X*X
100 GO TO 40
110 PRINT Z
120 END
```

Fig. 15

n	$(1 - 1/10^n)^{10^n}$
1	0.3486784401
2	0.3660323413
3	0.3676954248
4	0.3678610463
5	0.3678776027
6	0.3678792580
7	0.3678794134
8	0.3678792814
9	0.3678792327
10	0.3678794187
11	0.3678792328
12	1.0000000000

Fig. 16

7. a) 8. b) P ← AA; P ← AP; P ← PP; P ← PP; P ← PP; P ← P/A.

c) X ← AA; Y ← XA; Z ← XY; Z ← ZZ; Z ← ZZ; Z ← ZY.

8. a) 15 b) P ← AA; P ← PP; X ← PP; Y ← XX; P ← YY;

P ← PP; P ← PP; P ← PP; P ← PP; P ← PP; D ← XY; P ← P/D.

The program depends on the identity $A^{1000} = A^{1024} / (A^8 . A^{16})$.

c) We give the sequence of exponents used to calculate A^{1000}:

1, 2, 4, 5, 10, 20, 40, 80, 120, 125, 250, 500, 1000.

9. a) For A^{77} the exponent sequence is 1, 2, 4, 8, 9, 17, 34, 43, 77.

b) For A^{170} the exponent sequence is 1, 2, 3, 5, 10, 20, 40, 80, 85, 170.

10. 6, 4 and 14. In general this method of calculating a^n requires $\log_2 n + b(n) - 1$ multiplications, where $b(n)$ is the number of digits '1' in the binary representation of n. By comparison, Fig. 1.30 requires $\log_2 n + b(n)$ multiplications.

1.3.3

1. For $x = \frac{n}{2}$, where n is a whole number.

2. For 0, 1, 4, 9, . . ., n^2, . . .

6. $\frac{[10^d x + 0.5]}{10^d}$ rounds x to d decimal places.

7. c) Saturday. d) – f) are dates of three outbreaks of war (Russia, Pearl Harbour, Korea.) So they are Sundays.

8. a)

It is easily shown that $f(N + 1) - f(N) < 2$, so that no two adjacent adjacent integers are omitted. Fig. 17 prints the values omitted, X and Y are two successive values of the function. The numbers omitted are precisely the squares. The proof is elementary, though not trivial.

8. b) The triangular numbers 1, 3, 6, . . ., $\frac{n(n + 1)}{2}$, are omitted.

```
X ← 0
N ← 1
Y ← [N + √N + 0.5]
IF Y = X + 1
PRT X + 1
X ← Y
N ← N + 1
```

Fig. 17

10. The sequence of digits is reversed. e.g. N = 1984 becomes M = 4891.

1.3.4

1. a) $\frac{1}{\sqrt{x+1} + \sqrt{x}}$ b) $\frac{-1}{x(x+1)}$ c) $\frac{2}{x(x^2 - 1)}$

d) $\frac{1}{(\sqrt[4]{x+1} + \sqrt[4]{x})(\sqrt{x+1} + \sqrt{x})}$

2. $y = \sqrt[7]{x} \Leftrightarrow y^7 = x \Leftrightarrow y^8 = xy \Leftrightarrow y = \sqrt{\sqrt{\sqrt{xy}}}$ for $y > 0$.

Fig. 18 shows the program.

3. $y = \sqrt[9]{x} \Leftrightarrow y^9 = x \Leftrightarrow y^8 = \frac{x}{y} \Leftrightarrow y = \sqrt{\sqrt{\sqrt{\frac{x}{y}}}}$. The program in Fig. 19 has the convergence factor $\frac{1}{8}$.

4. See Fig. 20.

5. $\sqrt{i} = (1 + i)\sqrt{0.5}$

6. a) See Fig. 21.

b) $\lim\limits_{n \to \infty} x_n = \lim\limits_{n \to \infty} y_n = \sqrt{x_0 y_0}$

c) $x_{n+1} y_{n+1} = x_n y_n = x_0 y_0 \Rightarrow y_n = x_0 y_0 / x_n \Rightarrow x_{n+1} = \frac{1}{2}(x_n + x_0 y_0 / x_n)$

$\lim\limits_{n \to \infty} x_n = \sqrt{x_0 y_0} \Rightarrow \lim\limits_{n \to \infty} y_n = \sqrt{x_0 y_0}.$

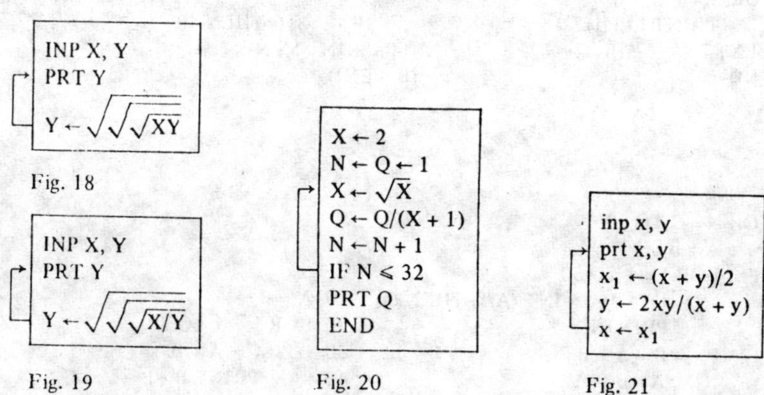

Fig. 18

Fig. 19

Fig. 20

Fig. 21

1.3.5

1. First calculate $\log_b \frac{1}{a}$. Then $\log_b a = -\log_b \frac{1}{a}$.

2. This is Fig. 1.38 for b = 5.

4. See Fig. 22.

7. Use Figs 1.38, 1.42. $\lg \pi = 0.4971498727$, $\lg e = 0.4342944819$,

$\ln 10 = 2.302585093$, $\ln \pi = 1.144729886$, $e^{\frac{\pi}{4}} = 2.193280051$,

$2^{\sqrt{2}} = 2.665144143$, $\pi^{\pi} = 36.46215961$.

8. $\lg 2 = 0.3010299957$, $\lg 3 = 0.4771212547$, $\lg 5 = 0.6989700044$.

1.3.6

6. The algorithm converges to $\frac{\sqrt{5} - 1}{2}$. The starting value $\frac{\sqrt{5} - 1}{2}$ consequently gives a constant sequence. The starting value $\frac{\sqrt{2}}{2}$ gives a constant sequence from its second member $\sqrt{2} - 1$. In both cases rounding errors soon distort the constant sequence.

```
10 A = 2
20 FOR I = 1 TO 600
30     Z = 1
40     IF A < 5 THEN 80
50     A = A/5
60     Z = Z + 1
70     GO TO 40
80     B (Z) = B (Z) + 1
90     A = A * A * A
100    A = A * A
110 NEXT I
120 FOR I = 1 TO 6
130    PRINT I, B (I)
140 NEXT I
150 END
```

Fig. 22

```
10 INPUT M
20 N = 2
30 A = B = 1
40 B = A + B
50 A = B - A
60 B = B - 10 ↑ M * INT (B/10 ↑ M)
70 N = N + 1
80 IF B < > 0 THEN 40
90 PRINT M, N
100 END
```

Fig. 23

```
10 FOR N = 2 TO 20
20     A = B = 1
30     FOR I = 1 TO 50
40         IF A/I < > INT (A/I) THEN 60
50         PRINT I;
60         B = A + B
70         A = B - A
80     NEXT I
90     PRINT
100 NEXT N
110 END
```

Fig. 24

```
10 A = B = 1
20 FOR I = 1 TO 50
30     C = A + B
40     PRINT B * B - A * C
50     B = A + B
60     A = B - A
70 NEXT I
80 END
```

Fig. 25

1.4

2. $L(10^n) = \text{LCM}\left\{3 \cdot 2^{n-1},\ 2^2 \cdot 5^n\right\} = \begin{cases} 12 \cdot 5^n & \text{for } n \leqslant 3 \\ 15 \cdot 10^{n-1} & \text{for } n \geqslant 3 \end{cases}$

3. See Fig. 23. For m = 1, 2, 3, 4 one finds n = 15, 150, 750, 7500.

4. See Fig. 24. With the notation "a | b" for "a divides b" one has, for example, $2|F_n \Leftrightarrow 3|n$, $3|F_n \Leftrightarrow 4|n$, $4|F_n \Leftrightarrow 6|n$, $5|F_n \Leftrightarrow 5|n$, $7|F_n \Leftrightarrow 8|n$ and so on.

7. See Fig. 25. The conjecture $F_n^2 - F_{n-1}F_{n+1} = (-1)^{n+1}$ may be proved by induction.

8. a) Induction gives $L_n^2 - L_{n-1}L_{n+1} = 5(-1)^{n+1}$

b) $\lim_{n\to\infty} L_{n+1}/L_n = \varphi = \frac{\sqrt{5}+1}{2}$.

9. See Fig. 26. The digit n has the relative frequency $h_n = \lg(1 + \frac{1}{n})$.

As soon as $A \geq 10$, A and B are divided by 10. That does not alter the first digit E. Also we have that $E = [A]$ and avoid overflow.

```
10 A = B = 1
20 FOR I = 1 TO 10000
30      E = INT (A)
40      R (E) = R (E) + 1
50      B = A + B
60      A = B – A
70      IF A < 10 THEN 100
80      B = B/10
90      A= A/10
100 NEXT I
110 FOR I = 1 TO 9
120      PRINT I, R (I)
130 NEXT I
140 END
```

Fig. 26

```
10 INPUT Z, B
20 Q = 0
30 Q = Q + Z – B * INT (Z/B)
40 Z = INT (Z/B)
50 IF Z > 0 THEN 30
60 PRINT Q
70 END
```

Fig. 27

2.1

2. See Fig. 27

3. a) See Fig. 28. b) A multiple of 3 is transformed into a multiple of 3 under this operation. A = 1899 is the largest multiple of 3 which is increased by this algorithm. So we need only consider the multiples of 3, $A \leq 1899$. One can easily think of further short-cuts.

4. See Fig. 29.

```
10 INPUT A
20 K = 0
30 PRINT A;
40 IF A = 153 THEN 100
50 Z = A – 10 * INT (A/10)
60 K = K + Z ↑ 3
70 A = INT (A/10)
80 IF A > 0 THEN 50
85 A = K
90 GO TO 20
100 END
```

Fig. 28

```
10 INPUT X, Y
20 Z = 1
30 IF Y = 0 THEN 90
40 IF Y/2 = INT (Y/2) THEN 60
50 Z = Z * X
60 X = X * X
70 Y = INT (Y/2)
80 GO TO 30
90 PRINT Z
100 END
```

Fig. 29

5. See Fig. 30.

6. To Fig. 2.10, the rows of Fig. 31 must be added.

7. See Fig. 32. The game is favourable to Abel. It can be shown that $A \sqcap B = 1$ with probability $p \approx \frac{6}{\pi^2} \approx 0{\cdot}6079$. See [5], pp. 201-212.

8. See Fig. 33a and 33b.

9. Use $A \sqcup B = (AB)/(A \sqcap B)$.

```
10 X (1) = X (2) = X (3) = X (4) = 0
20 FOR I  = 1 TO 1000
30      A = INT (1 E6 * RND) + 1
40      B = INT (1 E6 * RND) + 1
50      Q = B/A
60      IF Q > = 1 THEN 80
70      Q = 1/Q
80      IF Q > = 5 THEN 100
85      Q = INT (Q)
90      X (Q) = X (Q) + 1
100 NEXT I
110 PRINT X (1), X (2), X (3), X (4)
120 END
```

Fig. 30

```
10 C = 1E + 6
20 FOR I = 1 TO 1000
30      A= INT (C * RND) + 1
40      B= INT (C * RND) + 1
50      A= A - B * INT (A/B)
60      IF A = 0 THEN 90
70      B = B - A * INT (B/A)
80      IF B < > 0 THEN 50
90      IF A + B < > 1 THEN 110
100     G = G + 1
110 NEXT I
120 PRINT G
130 END
```

Fig. 32

```
 5 D = 0
45 D = D + 1
65 D = D + 1
85 PRINT D
```

Fig. 31

```
10 A = 2
20 B = 1
30 PRINT A * A - B * B, 2 * A * B, A * A + B * B
40 B = B - 2
50 IF B > 0 THEN 100
60 B = A
70 A = A + 1
80 IF A > 10 THEN 180
90 GO TO 30
100 P = A
110 Q = B
120 P = P - Q * INT (P/Q)
130 IF  P = 0  THEN 160
140 Q = Q - P * INT (Q/P)
150 IF Q < > 0  THEN 120
160 IF P + Q = 1 THEN 30
170 GO TO 40
180 END
```

Fig. 33a

X	Y	Z
3	4	5
5	12	13
7	24	25
15	8	17
9	40	41
21	20	29
11	60	61
35	12	37
13	84	85
33	56	65
45	28	53
15	112	113
39	80	89
55	48	73
63	16	65
17	144	145
65	72	97
77	36	85
19	180	181
51	140	149
91	60	109
99	20	101

Fig. 33 b

2.3

1. a) $d = 7$, $x = -3$, $y = 5$ b) $d = 11$, $x = 3$, $y = -7$

   c) $d = 1$, $x = -6$, $y = 55$

   d) $d = 1$, $x = 34 = F_9$, $y = -55 = -F_{10}$, i.e. $F_{12}F_9 - F_{11}F_{10} = 1$

   e) $d = 1$, $x = -55$, $y = 89 = F_{11}$, i.e. $F_{13}F_{10} - F_{12}F_{11} = -1$

   f) $d = 1$, $x = 89$, $y = -144 = -F_{12}$, i.e. $F_{14}F_{11} - F_{13}F_{12} = 1$ etc.

   The conjecture is $F_nF_{n-3} - F_{n-1}F_{n-2} = (-1)^n$.

2. $x = 31$, $y = 63$.

2.4

2. See Fig. 34. 135, 127, 120, 106 prime numbers.
3. See Fig. 35.
4. See Fig. 36. P = previous prime, L = current maximum gap-length.
5. See Fig. 37.

```
 10 INPUT M, N
 20 P = 0
 30 FOR A = M TO N STEP 2
 40       FOR B = 3 TO SQR (A) STEP 2
 50             IF A/B = INT (A/B) THEN 80
 60       NEXT B
 70       P = P + 1
 80 NEXT A
 90 PRINT P
100 END
```

Fig. 34

```
 10 PRINT 1;
 20 P = 3
 30 FOR A = 5 TO 1000 STEP 2
 40       FOR B = 3 TO SQR (A) STEP 2
 50             IF A/B = INT (A/B) THEN 90
 60       NEXT B
 70       PRINT A - P;
 80       P = A
 90 NEXT A
100 END
```

Fig. 35

```
 10 INPUT M, N, P
 20 L = 2
 30 FOR A = P + 2 TO N STEP 2
 40       FOR B = 3 TO SQR (A) STEP 2
 50             IF A/B = INT (A/B) THEN 100
 60       NEXT B
 70       IF A - P <= L THEN 90
 80       L = A - P
 90       P = A
100 NEXT A
110 PRINT L
120 END
```

Fig. 36

```
 10 P = 3
 20 A = 5
 30 FOR B = 3 TO SQR (A) STEP 2
 40       IF A/B = INT (A/B) THEN 80
 50 NEXT B
 60 IF A - P = 16 THEN 100
 70 P = A
 80 A = A + 2
 90 GO TO 30
100 PRINT P, A
110 END
```

Fig. 37

6. $123456789 = 3^2 . 3607 . 3803$, $987654321 = 3^2 . 17^2 . 379721$,
$2^{32} + 1 = 641 . 6700417$, $1264460 = 2^2 . 5 . 17 . 3719$,
$81128632 = 2^3 . 13 . 19 . 41057$, 600 000 017 is a prime.

7. See Fig. 38.

9. See Fig. 39, 40a, 40b.

```
 10 INPUT N
 20 FOR D = 2 TO SQR (N)
 30       IF N/D = INT (N/D) THEN 70
 40 NEXT D
 50 PRINT N
 60 GO TO 100
 70 PRINT D;
 80 N = N/D
 90 GO TO 20
100 END
```

Fig. 38

n	h (n)	q (n)
10	7	0.7
20	13	0.65
30	19	0.63333
40	26	0.65
50	31	0.62
60	37	0.61667

Fig. 39

```
10 DIM X (2000)
20 H = 0
30 FOR I = 1 TO 2000
40       X (I) = 0
50 NEXT I
60 READ S
70 FOR  I = S  TO 2000 STEP S
80       X (I) = 1
90 NEXT I
100 IF S < 1849 THEN 60
110 FOR  I = 1 TO 2000
120       H = H + 1 - X (I)
130       IF I/100 < > INT (I/100) THEN 150
140       PRINT I, H, H/I
150 NEXT I
160 DATA 4, 9, 25, 49, 121, 169, 289, 361
170 DATA 529, 841, 961, 1369, 1681, 1849
180 END
```

Fig. 40 a

n	h (n)	q (n)
100	61	0.61
200	122	0.61
300	183	0.61
400	243	0.6075
500	306	0.612
600	366	0.61
700	428	0.61143
800	489	0.61125
900	547	0.60778
1000	608	0.608
1100	667	0.60636
1200	730	0.60833
1300	792	0.60923
1400	854	0.61
1500	915	0.61
1600	977	0.61062
1700	1035	0.60882
1800	1096	0.60889
1900	1153	0.60684
2000	1215	0.6075

Fig. 40 b

2.5

1. See Fig. 41. a) If n is a prime, the period length, $p(n)$, is a divisor of $n - 1$. b) If $\varphi(n)$ is the number of positive integers $\leq n$ coprime with n, then $p(n)$ is a divisor of $\varphi(n)$.

```
10 FOR N = 3 TO 100 STEP 2
20       IF N MOD 5 = 0 THEN 70
30       R = 1;  P = 0
40             R = 10 * R MOD N;  P = P + 1
50             IF R < > 1 THEN 40
60       PRINT N; P,
70 NEXT N
80 END
```

Fig. 41

```
3 1   7 6   9 1   11 2   13 6   17 16   19 18   21 6   23 22   27 3   29 28   31 15
33 2   37 3   39 6   41 5   43 21   47 46   49 42   51 16   53 13   57 18   59 58
61 60   63 6   67 33   69 22   71 35   73 8   77 6   79 13   81 9   83 41   87 28
89 44   91 6   93 15   97 96   99 2
```

3. See Fig. 42.

```
10 INPUT N
20 R = 1
30 Q = INT (10 * R/N)
40       PRINT Q;
50       R = 10 * R - Q * N
60 IF R < > 1 THEN 30
70 END
```

Fig. 42

2.6

1. a) Fig. 43 uses the interchange instruction "= =" and the mod operation.

b) $\frac{F_{n+1}}{F_n} = [1; 1, 1, 1, \ldots, 1, \underline{2}]$. The proof follows from

$$\frac{F_{n+1}}{F_n} = \frac{F_n + F_{n-1}}{F_n} = 1 + \cfrac{1}{\cfrac{F_n}{F_{n-1}}}$$

2. $\left[\frac{1}{X}\right] = n \Leftrightarrow n \leq \frac{1}{X} < n + 1 \Leftrightarrow \frac{1}{n+1} < X \leq \frac{1}{n}$. Hence

$\overline{p}_n = \frac{1}{n} - \frac{1}{n+1} = \frac{1}{n(n+1)}$

```
10 INPUT A, B
20 PRINT INT (A/B);
30       A = A MOD B
40       A = = B
50 IF B <> 0 THEN 20
60 END
```

Fig. 43

```
10 PRINT 2;
20 FOR  N = 2  TO 1000
30         Y = N;  Z = 1,  X = 2
40         IF Y/2 = INT (Y/2) THEN 60
50         Z = Z * X MOD N
60         X = X * X MOD N;  Y = INT (Y/2)
70         IF Y <> 0 THEN 40
80         IF Z = 2 THEN PRINT N;
90 NEXT N
100 END
```

Fig. 44

2.7

1. See Fig. 44

2. n = 341, 561, 645.

4. a) n = 91, 121, 561, 671, 703, 949 b) n = 217, 561, 781.

3.2

1. $\pi = (2 - \frac{2}{9})^2 = \frac{256}{81} = 3.160493827$

3.3

1. n = 9

2. a) $\frac{\pi}{6}$, $\frac{\pi}{3}$, $\frac{\pi}{2}$ b) $\frac{\pi}{3}$, $\frac{\pi}{6}$, 0 c) $\frac{\pi}{4}$, $\frac{\pi}{3}$, $\frac{\pi}{2}$

3. $T_n = \frac{S_n}{c_n}$, $T_{n+1} = \frac{S_n}{c_{n+1}^2}$, $T_{n-1} = \frac{S_n c_n}{c_{n-1}}$,

$$\frac{T_{n+1} - T_n}{T_n - T_{n-1}} = \frac{c_{n-1}}{c_{n+1}^2} \; \frac{1 - c_n}{1 - c_{n-1}} \rightarrow \frac{1}{4}$$

4. a) Let n_i be the least n, so that $\alpha_1 + \alpha_2 + \ldots + \alpha_n \geqslant 2\pi i$.

Fig. 45 prints the following table:

i	1	2	3	4	5	6	7	8	9	10
n_i	17	54	110	186	281	396	532	686	801	1055

```
A ← 0
N ← I ← 1
A ← A + ATN (1/√N)
IF A < 2π
PRT I, N
I ← I + 1
A ← A − 2π
N ← N + 1
IF I ≤ 10
END
```

Fig. 45

5. a) $s_n = \sin \frac{\alpha}{2^n}$, $c_n = \cos \frac{\alpha}{2^n}$, $t_n = \tan \frac{\alpha}{2^n}$, $S_n = 2^{n+1}\sin \frac{\alpha}{2^n}$,

$T_n = 2^{n+1}\tan \frac{\alpha}{2^n}$

b) With $s_1 = 2 \sin x$, $s = 2 \sin \frac{x}{2}$, (1) gives $\sin x = 2 \sin \frac{x}{2} \cos \frac{x}{2}$.

With $S = 2 \sin x$, $s = 2 \sin \frac{x}{3}$, (2) gives $\sin x = 3 \sin \frac{x}{3} - 4 \sin^3 \frac{x}{3}$.

6. The algorithm has convergence factor $\frac{1}{4}$. Hence $n = 16$.

3.5

4. Convergence factor $\frac{1}{4}$.

6. Convergence factor $\frac{1}{4}$.

7b. See Fig. 46.

8b. See Fig. 47.

10. $s_n = \sin h \frac{t}{2^n}$, $c_n = \cos h \frac{t}{2^n}$, $t_n = \tan h \frac{t}{2^n}$, $S_n = 2^n \sin h \frac{t}{2^n}$,

$T_n = 2^n \tan h \frac{t}{2^n}$.

11. Convergence factor $\frac{1}{16}$.

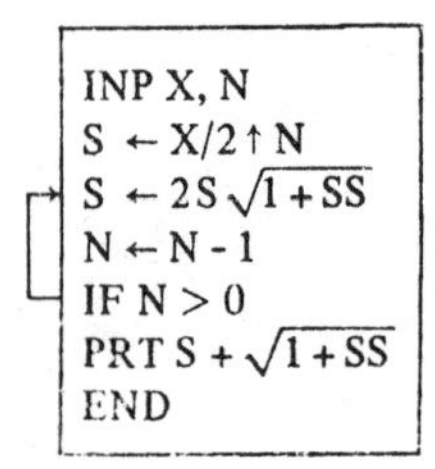

Fig. 46

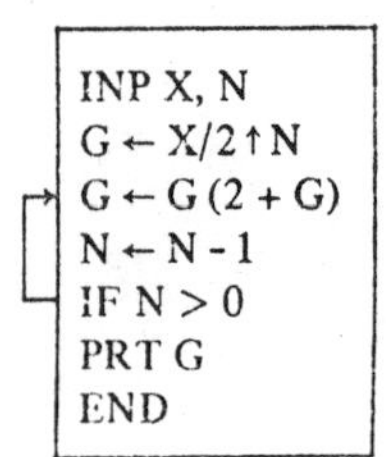

Fig. 47

3.8

2. See Fig. 48 - 50. a) 3.141092654 b) 3.142092904 c) 3.141592528

3. See Fig. 51.

4. 3.140807746. Fig. 52 depends on the transformation $\frac{2}{\pi} = \prod_{i=1}^{1000} (1 - \frac{1}{4i^2})$.

5. a) 3.140638056 b) 3.141592652

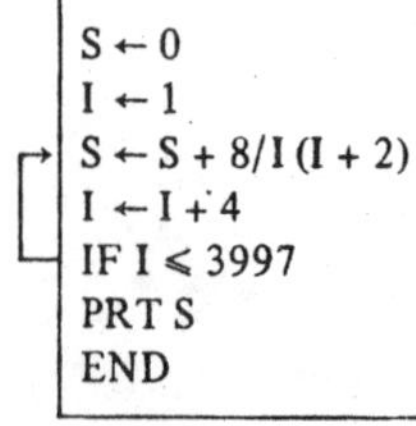

Fig. 48

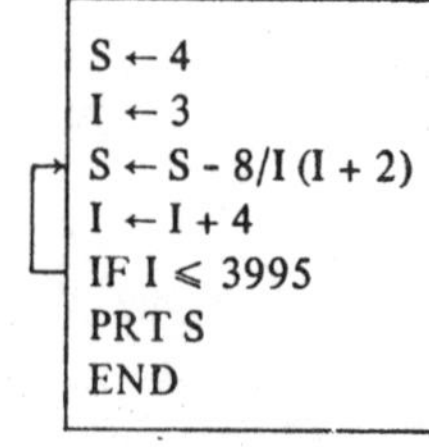

Fig. 49

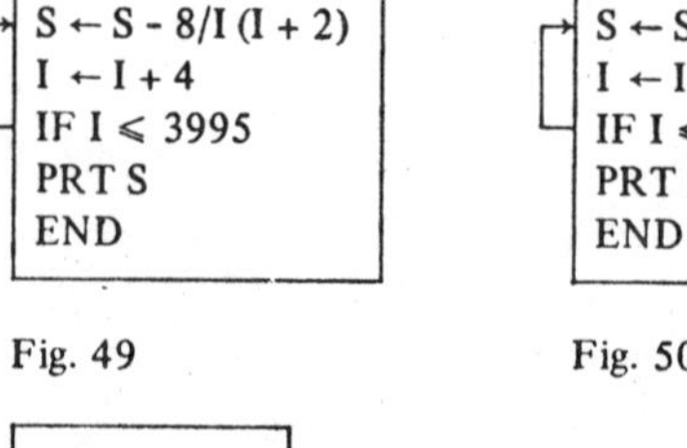

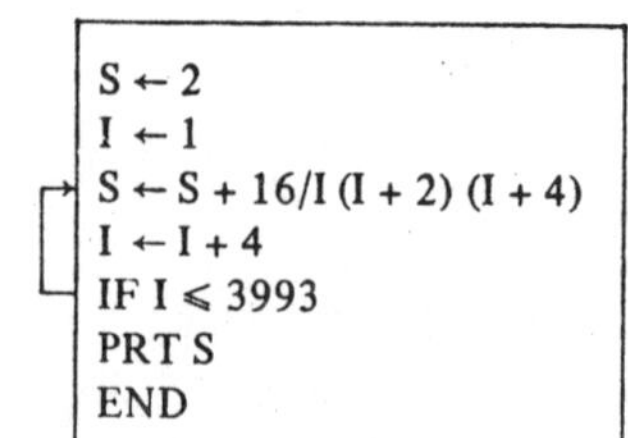

Fig. 50

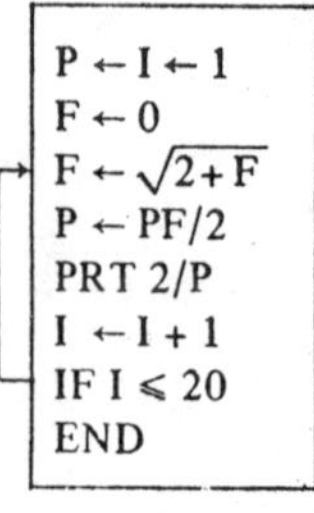

Fig. 51

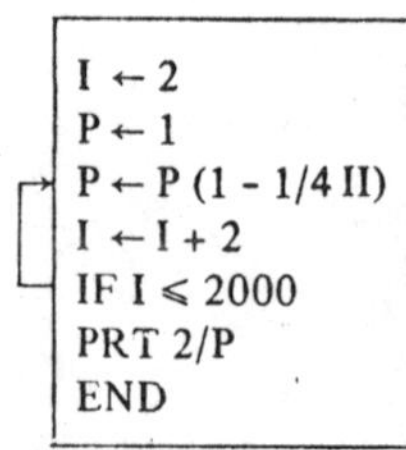

Fig. 52

3.9

1. The two needles make the angles a and $a_1 = a + \frac{\pi}{2}$ with the Northerly direction in Fig. 53. The program in Fig. 55 is intelligible without commentary.

2. The midpoint (x, y) of the needle is chosen randomly in the unit square. Then the orientation of the needle is chosen randomly between 0 and π. Then whether the needle intersects a side of the square, or one of the sides produced, is determined. See Fig. 54 and 56.

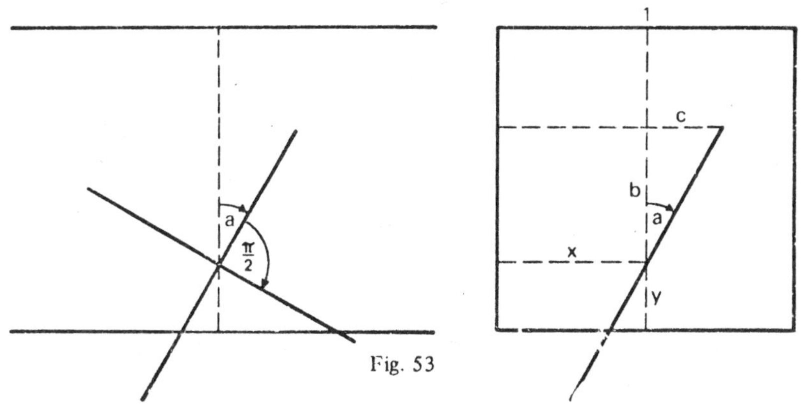

Fig. 53

Fig. 54

```
10 S = 0
20 FOR W = 1 TO 1000
30        Y = RND; A = PI * RND
40        B = COS (A)/2;  B1 = COS (A + PI/2)/2
50        IF INT (Y - B) <> INT (Y + B) THEN S = S + 1
60        IF INT (Y - B1) <> INT (Y + B1) THEN S = S + 1
70 NEXT W
80 PRINT 4 * W/S
90 END
```

Fig. 55

```
10 S = 0
20 FOR W = 1 TO 1000
30        X = RND; Y = RND; A = PI * RND
40        B = COS (A)/2; C = SIN (A)/2
50        IF INT (Y - B) <> INT (Y + B) OR INT (X - C) <> INT (X + C) THEN S = S + 1
60 NEXT W
70 PRINT 3 * W/S
80 END
```

Fig. 56

3. Fig. 57 shows the program for n = 4. Line 30 counts the points with positive coordinates which lie inside the sphere. This is $\frac{1}{16}$ of the whole sphere. Hence $\frac{16R}{1000}$ is an approximation to the volume of the sphere. The exact value is $\frac{\pi^2}{2}$.

```
10 R = 0
20 FOR I = 1 TO 1000
30        R = R + 1 - INT (RND↑2 + RND↑2 + RND↑2 + RND↑2)
40 NEXT I
50 PRINT 16 * R/1000
60 END
```

Fig. 57

5. Let the three random points be $P_1(x_1, y_1)$, $P_2(x_2, y_2)$, $P_3(x_3, y_3)$.

The squares of the sides of the triangle are $a = (x_1 - x_2)^2 + (y_1 - y_2)^2$, $b = (x_2 - x_3)^2 + (y_2 - y_3)^2$, $c = (x_1 - x_3)^2 + (y_1 - y_3)^2$. The triangle is obtuse if $a > b + c$ or $b > a + c$ or $c > a + b$ (Fig. 58 and 59).

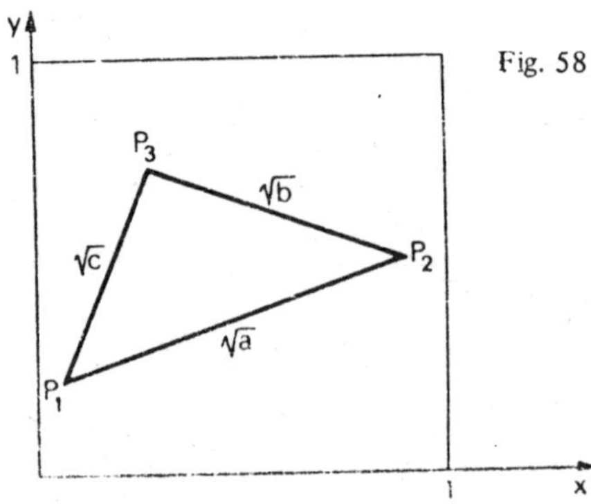

Fig. 58

```
10 FOR I = 1 TO 1000
20       X1 = RND; X2 = RND; X3 = RND; Y1 = RND; Y2 = RND; Y3 = RND
30       A = (X1 - X2)↑2 + (Y1 - Y2)↑2
40       B = (X2 - X3)↑2 + (Y2 - Y3)↑2
50       C = (X1 - X3)↑2 + (Y1 - Y3)↑2
60       IF A > B + C OR B > A + C OR C > A + B THEN S = S + 1
70 NEXT I
80 PRINT S/1000
90 END
```

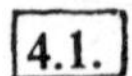

Fig. 59

6. $P = \frac{1}{4}$.

4.1.

1 a)

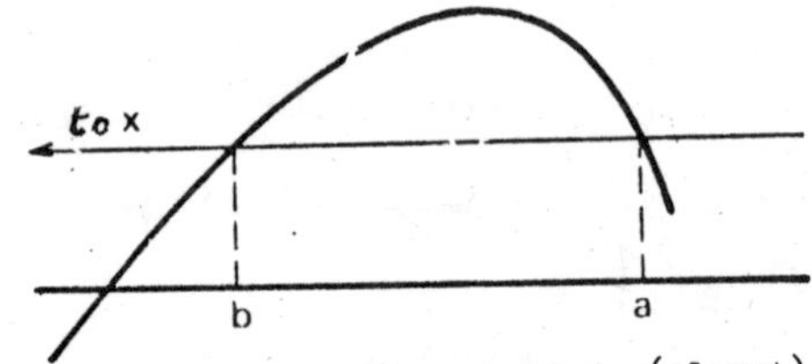

Fig. 60

The secant is (almost) parallel to the x-axis.

b)

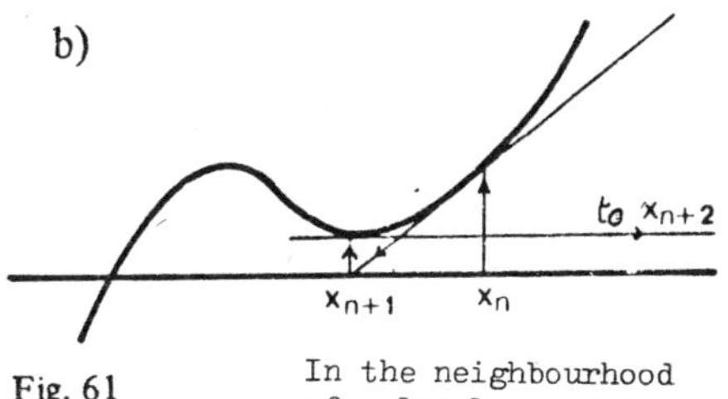

Fig. 61 In the neighbourhood of a local extremum.

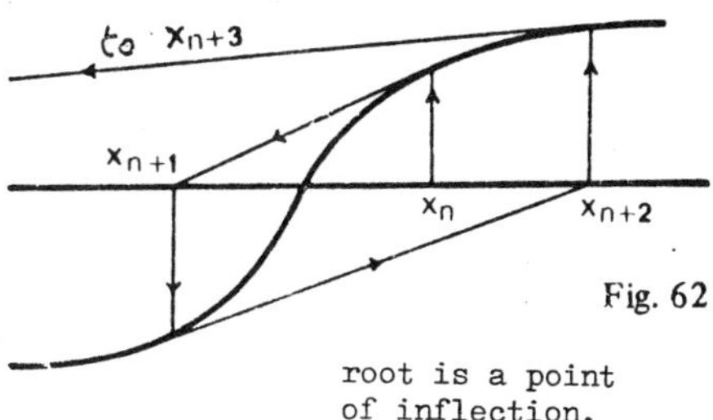

Fig. 62 root is a point of inflection.

3. $x = 2.195823345$

4. a) $x = 1.763222834$ b) $x = 2.154434690$ c) $x = 1.324717957$
   d) $x = 0.8767262154$ e) $x = 0.7390851332$ f) $x = -0.1999361022$

5. a) $x = 2.094551482$ b) $x = 1.584893192$ c) $x = 1.112775684$
   d) $x = 0.7390851332$ e) $x = 0.5671432904$

6. The initial values $x = 0$ and $x = 2$ give the two solutions.

   $x_1 = 0.3574029562$ and $x_2 = 2.153292364$.

7. $x_1 = 2 \sin 10^\circ = 0.3472963553$, $x_2 = 2 \sin 50^\circ = 1.532088886$,
   $x_3 = -2 \sin 70^\circ = -1.879385242$

10. This is the Newton iteration-sequence for $f(x) = x^m - a$. The graph of the iteration function $g(x) = ((m-1)x + \frac{a}{x^{m-1}})/m$ is similar to that of Fig. 4.19. One can see that, as in Example 4, $x_1, x_2, x_3 \dots$ is monotone decreasing toward $\sqrt[m]{a}$.

11. $|g(x) - g(y)| = \frac{|x-y|}{\sqrt{1+x} + \sqrt{1+y}} < \frac{|x-y|}{2}$ for $x, y > 0$.

    Fixed point $s = \frac{\sqrt{5} - 1}{2}$

12. $|g(x) - g(y)| = \frac{|x-y|}{xy} < |x - y|$ for $x > 1$ and $y > 1$.

    Fixed point $s = \frac{\sqrt{5} + 1}{2}$

13. The sequence converges for all $x_0 \geqslant -1$ to $s = \frac{1 + \sqrt{17}}{8}$

14. Show first that $x_{n+1} - \sqrt{10} = \frac{4 - \sqrt{10}}{x_n + 4}(x_n - \sqrt{10})$.

15. $s = \frac{1 + \sqrt{1 + 4a}}{2}$. For $a = m(m + 1)$, $s = m + 1$.

16. a) $s = \frac{\sqrt{5} - 1}{2}$ b) $s = \sqrt{2}$.

18. With $x_n = \frac{1 - \epsilon_n}{a}$, $x_{n+1} = \frac{1 - \epsilon_n^2}{a}$, i.e. $\epsilon_{n+1} = \epsilon_n^2$ or $\epsilon_n = \epsilon_0^{2^n}$.

From this it follows that the convergence is quadratic, to $\frac{1}{a}$, provided $-1 < \epsilon_0 < 1$, or $0 < x_o < \frac{2}{a}$.

19. The sequence converges for all $A < e^{\frac{1}{e}} = 1.444667861$. For $A = \sqrt{2}$ and $A = e^{\frac{1}{e}}$ the limits are 2 and e respectively. With the initial approximation $e^{\frac{1}{e}}$ the convergence towards e is very slow.

21. The Newton iteration function for the solution of the equation $\ln y - x =$ for y gives $g(y) = y(1 + x - \ln y)$. Fig. 4.19b prints the correspondin iteration-sequence for the starting value $y = 1$. The program works correctly for $x_o > -1$.

24. My computer gives for $n = 10, 100, 1000, 10000$ the values

1, 0·9999999992, 0·9999999902, 0·9999990004.

Fig. 63 shows the corresponding program.

```
10 INPUT N
20 X = 1; Y = 0
30 FOR I = 1 TO N
40      A = (3 * X - 4 * Y)/5; B = (4 * X + 3 * Y)/5
50      X = A; Y = B
60 NEXT I
70 PRINT X * X + Y * Y
80 END
```

Fig. 63

```
x ← a
m ← f (x)
x ← x + h
if m ≤ f (x)
h ← - h/10
if 10 | h | > ε
prt x, m
end
```

Fig. 64

4.2

1. See Fig. 64

2. The fourth line should read if $m \geq f(x)$.

a) $x = \sqrt[3]{4} = 1.587401052$ b) $x = e^{-1} = 0.3678794412$

3. In Fig. 65 x converges to the abscissa of B. The mistake can be corrected by going two steps back in 3. ($x \leftarrow x - 2h$).

4. a) $x = \frac{3\pi}{4} = 2{\cdot}35619449$ b) $x = 1$

5. See Fig. 66.

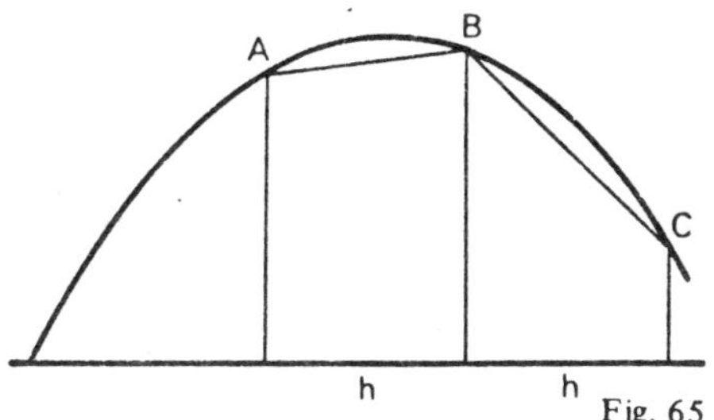

Fig. 65

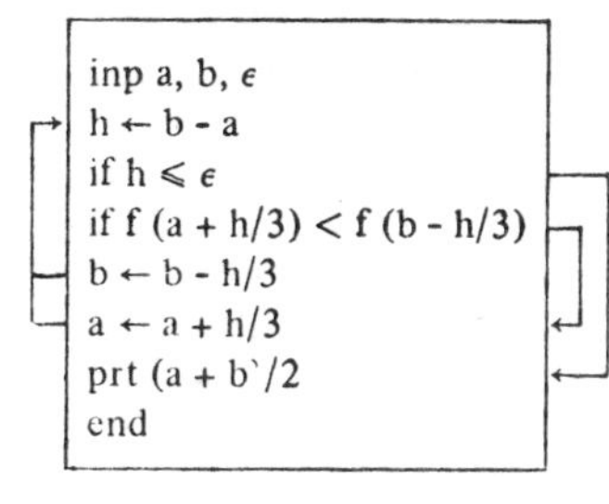

Fig. 66

4.3.2

1. a) 0.5939941503 b) 3.059116540 c) 0.7468241328

2. The evaluation of the first integral is difficult because the curve has a vertical tangent at $x = 1$. The arc of the curve in the neighbourhood of $x = 1$ cannot be closely approximated by a parabola. By contrast the second method speedily gives exact values for the integral. One obtains $I = 8{\cdot}708149356$.

3. a) See Fig. 67

```
 10 INPUT A, B
 20 FOR I = 0 TO 12
 30     M = 0
 40     H = (B - A)/2↑I
 50     FOR X = A + H/2 TO B STEP H
 60         M = M + 1/LOG (X)
 70     NEXT X
 80     M = M * H
 90     PRINT I, M
100 NEXT I
110 END
```

Fig. 67

4.4.1

1. a) Fig. 68 b) Fig. 69 c) $e - e_{15} < \frac{17}{16} \cdot \frac{1}{16!} < 5{\cdot}08 \cdot 10^{-14}$

2. Fig. 70

3. Fig. 71 gives a table for a_n and Fig. 16 for b_n. In both cases the monotone growth is disturbed by rounding error for $n > 10^7$.

4. Fig. 72

5. Fig. 73

6. Fig. 74

7. Fig. 75

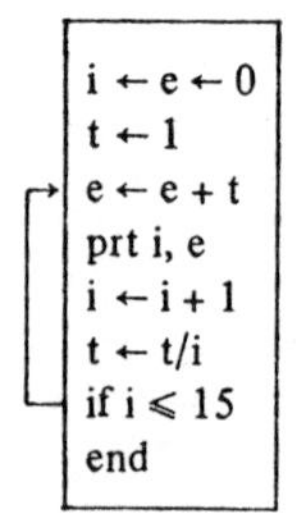

Fig. 68

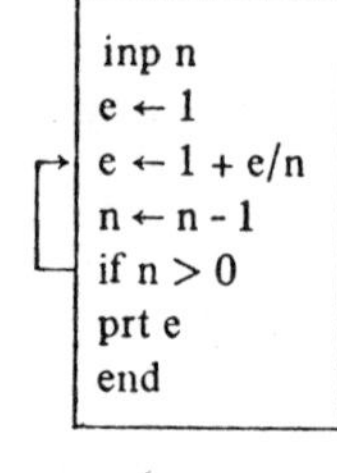

Fig. 69

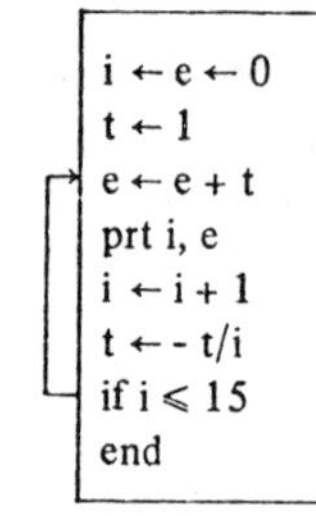

Fig. 70

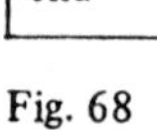

n	$(1 + \frac{1}{10^n})^{10^n}$
1	2.593742460
2	2.704813829
3	2.716923927
4	2.718145849
5	2.718268163
6	2.718272755
7	2.718276320
8	2.718271543
9	2.718271795
10	2.718275492
11	2.718270349
12	1.000000000

Fig. 71

```
i ← s ← 0
t ← 1
s ← s + t
i ← i + 1
t ← tx/i
if | t/s | > f
prt x, s
end
```

Fig. 72

n	$(1 + \frac{1}{n} + \frac{1}{2n^2})^n$	$(1 + \frac{1}{n - 0,5})^n$
10	2.714080847	2.720551414
10^2	2.718236860	2.718304483
10^3	2.718281366	2.718282073
10^4	2.718281828	2.718281828

Fig. 73

n	e_n
1	2.732050808
2	2.719199680
4	2.718340389
8	2.718285508
16	2.718282059
32	2.718281842
64	2.718281832
128	2.718281828

Fig. 74

n	e_n
1	2.714285714
2	2.718042367
4	2.718267026
8	2.718280906
16	2.718281771
32	2.718281824
64	2.718281830
128	2.718281828

Fig. 75

8. If we write in (X), $n = 1000$ and $t = \frac{1}{2}$, then we obtain $\sqrt{e} \approx 1.648721264$. The exact value is $e = 1.648721271$.

9. Put $t = -1$ in (X), obtaining $c_{1000} = 0.3678794423$.

10. a) $e^{1/2n} - e^{-1/2n}$ b) $(e^{1/2n} - e^{-1/2n}) /2n$

c) Put $u_n = e^{1/2n}$ and solve $u_n - 1/u_n = (u_n + 1/u_n) /2n$ for u_n.

11. Fig. 76

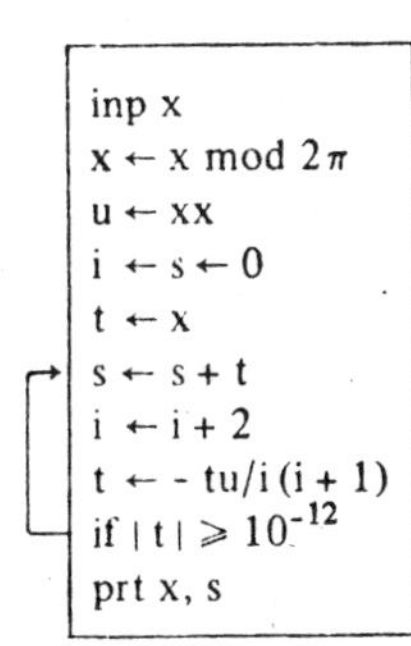

Fig. 76

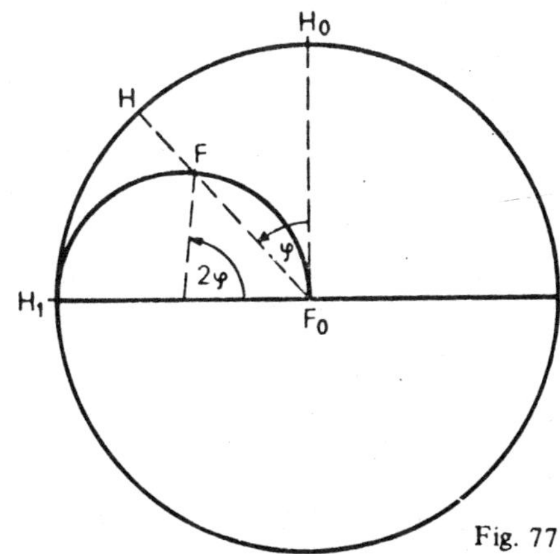

Fig. 77

4.4.3

3. b) Show that in Fig. 77 $\widehat{H_0H} = \widehat{F_0F}$.

4. a) 0.202594 b) 0.197469 c) 0.197200 d) 0.202301

About 20% of the population never hear the rumour. Surprisingly this result is almost independent of p and n.

5.1

2. b) The rows of the Pascal triangle are numbered 0, 1, 2, 3, ...

Rows numbered $2^n - 1$ contain only 'ones' and rows numbered 2^n contain only 'zeros', where $n = 0, 1, 2, \ldots$

5.2

1. A US-Dollar can be changed in 292 ways.

2. A Swiss franc can be changed in 4562 ways.

3. 4332 solutions.

```
 10 N = 100
 20 M = N/EXP(1)
 30 P = 1/SQR(2 * PI * N)
 40 Q = 1 + 1/(12 * N)
 50 R = Q + 1/(288 * N * N)
 60 FOR I = 100 TO 1 STEP - 1
 70      P = P * I/M
 80 NEXT I
 90 PRINT P, P/Q, P/R
100 END
```

Fig. 78

5.3

1. b) Fig. 78 prints the three quotients for n = 100. The results are

   1.000833679, 1.000000345, 0.9999999981

2. The programs in Fig. 5.11 and 5.16 should be run successively. See also the solution of the next exercise.

3. See Fig. 79. The program consists of Fig. 5.11 and a slight modification of Fig. 5.16. The variable Z counts the cycles.

4. The desired program is obtained bycompleting Fig. 79.

```
 10 DIM L (100)
 20 FOR I = 1 TO 100 L (I) = I
 30 FOR I = 100 TO 2 STEP - 1
 40      K = INT (I * RND) + 1
 50      L (I) = = L (K)
 60 NEXT I
 70 I = 1; Z = 0
 80 E = I
 90 K = I; I = L (I); L (K) = - L (K)
100 IF I <> E THEN 90
110 Z = Z + 1
120 FOR I = 1 TO 100
130      IF L (I) > 1 THEN 80
140 NEXT I
150 PRINT Z
160 END
```

Fig. 79

```
10 INPUT T, S
20 Q = I = 1
30 IF 1 - Q > S THEN 70
40 Q = Q * (T - I)/T
50 I = I + 1
60 GO TO 30
70 PRINT I, 1 - Q
80 END
```

Fig. 80

5.4

1. For n = 23 we have $p_n = 0.5073$.

2. a) $p_{41} = 0.90315$ b) $p_{57} = 0.99012$ c) $p_{70} = 0.99916$

   d) $p_{80} = 0.99991$. Use Fig. 80 with T = 365.

3. a) n = 38 b) n = 68 c) n = 95 d) n = 116

   Use Fig. 80 with T = 1000 and S = 0.5, 0.9, 0.99, 0.999.

4. Using the program in Fig. 80 one quickly finds by trial that

$$1713 \leq x \leq 1783.$$

5. See Fig. 81, where $s = b(0) + \ldots + b(a)$

```
10 INPUT A, N, P
20 Q = 1 - P
30 S = B = Q↑N
40 FOR X = 1 TO A
50        B = B * P * (N - X + 1)/(Q * X)
60        S = S + B
70 NEXT X
80 PRINT A, N, P, S
90 END
```

Fig. 81

6. a) For $a = 39$, $n = 100$, $p = \frac{1}{2}$, Fig. 81 gives $s = 0.0176$

b) The probability of $|x - 50| > 10$ is twice as big as that of $x < 40$ i.e. $s = 0.0352$.

7. For $n = 600$, $p = \frac{1}{6}$ and $a = 80$ and $a = 119$ respectively we obtain $s_1 = b(0) + \ldots + b(80) \approx 0.0144676$,

$s_2 = b(0) + \ldots + b(119) \approx 0.981989$.

Thus $s_3 = b(120) + \ldots + b(600) = 1 - s_2 \approx 0.0180$. Hence the desired probability is $s_1 + s_3 \approx 0.0325$.

The input $(x - a, n, q) = (480, 600, \frac{5}{6})$ led to underflow since $(\frac{1}{6})^{600}$ is too small for the computer to handle.

9. a) See Fig. 82.

c) For $n = 200$ we obtain $1 - \frac{196}{10^9} \approx 1 + \frac{b}{40000}$, i.e., $b \approx \frac{1}{128}$.

Hence

$$b_n \sim \frac{1}{\sqrt{\pi n}} \left(1 - \frac{1}{8n}\right)\left(1 + \frac{1}{128n^2}\right)$$

or, more simply

$$\boxed{b_n \sim \frac{1}{\sqrt{\pi n}} \left(1 - \frac{1}{8n} + \frac{1}{128n^2}\right)}$$

```
A ← 1
B ← 0.5
IF A/100 ≠ [A/100]
PRT A, B √(πA) / (1 − 1/(8A))
A ← A + 1
B ← B (2A − 1)/(2A)
IF A ≤ 1000
END
```

Fig. 82a

n	c_n
100	1.000000787
200	1.000000196
300	1.000000087
400	1.000000048
500	1.000000030
600	1.000000020
700	1.000000014
800	1.000000010
900	1.000000007
1000	1.000000005

Fig. 82b

Note that in Fig. 82b the transition from 100 to 200 or from 200 to 400 reduces the deviation from 1 almost exactly by a factor of four. This is not the case for the transitions from 400 to 800 or 500 to 1000, because of rounding error.

6.2

1. Fig. 83 arises by modification and completion of Fig. 6.8. The probability of neighbours is 0·495. See [4], pp. 44 and 166.

4. Fig. 84 shows the program for $n = 1000$. Z is the total number of moves. S counts the wins of Black. R is a random number from $\{1,2\}$. The colourings of the two circles are stored in X(1) and X(2). White stores +1 if he claims the circle and Black stores −1. If X(1) + X(2) is 2 or −2 then White or Black respectively has won. The average game length and the probabilities of winning are given in [4], p. 38.

5. Fig. 85 is completely analogous to Fig. 6.15. C is the value of χ^2.

```
10  DIM L (49)
20  A = 0
30  FOR J = 1 TO 100
40        FOR  I = 1 TO 49
50              L (I) = I
60        NEXT I
70        FOR K = 1 TO 6
80              I  = INT (49 * RND) + 1
90              IF L (I) < 0 THEN 80
100             L (I) = - 1
110       NEXT K
120       FOR I = 1 TO 48
130             IF L (I) + L (I + 1) = - 2 THEN 160
140       NEXT I
150       GO TO 170
160       A = A + 1
170 NEXT J
180 PRINT A/100
190 END
```

Fig. 83

```
10  Z = S = 0
20  FOR I = 1 TO 1000
25        X (1) = X (2) = 0
30        R = INT (2 * RND) + 1
40        Z = Z + 1
50        X (R) = 1
60        IF X (1) + X (2) = 2 THEN 120
70        R = INT (2 * RND) + 1
80        Z = Z + 1
90        X (R) = - 1
100       IF X (1) + X (2) <> - 2 THEN 30
110       S = S + 1
120 NEXT I
130 PRINT Z/1000, S/1000
140 END
```

Fig. 84

```
10  C = 0
20  FOR I = 1 TO 6
30        B (I) = 0
40  NEXT I
50  FOR I = 1 TO 600
60        D = INT (6 * RND) + 1
70        B (D) = B (D) + 1
80  NEXT I
90  FOR  I = 1 TO 6
100       PRINT B (I);
110       C = C + (B (I) - 100)↑2/100
120 NEXT I
130 PRINT
140 PRINT C
150 END
```

Fig. 85

6. Fig. 86 shows a program equivalent to Fig. 6.25. Lines 10 to 50 of this program replace lines 30 to 70 in Fig. 6.26, and lines 10 to 40 replace lines 70 to 100 in Fig. 6.27.

```
10  Y (1) = Y (2) = Y (3) = I = 0
20  D = INT (3 RND) + 1
30  Y (D) = 1 - Y (D); I = I + 1
40  IF Y (1) + Y (2) + Y (3) < 3 THEN 20
50  PRINT I
60  END
```

Fig. 86

7. See Fig. 87, where G is the weight of the latest fish caught.

We conjecture that both stopping-rules gives the same distribution for X. The proof is given in [5], pp. 93-97.

8. In Fig. 88, E is the weight of the first fish. The frequency of X >10 is counted in R(1).

```
10  S = 0
20  FOR X = 2 TO 10 R (X) = 0 NEXT X
30  FOR I = 1 TO 1000
40      X = G = 0
50      G = G + RND; X = X + 1
60      IF G < 1 THEN 50
70      R (X) = R (X) + 1; S = S + X
80  NEXT I
90  FOR X = 2 TO 10
100     IF R (X) < >0 THEN PRINT X, R (X)/1000
110 NEXT X
120 PRINT
130 PRINT "MEAN NUMBER CAUGHT  =";S/1000
140 END
```

Fig. 87

X	R (X)
2	479
3	350
4	127
5	36
6	6
7	2

MEAN NUMBER CAUGHT = 2.746

```
10  S = 0
20  FOR X = 1 TO 10 R (X) = 0 NEXT X
30  FOR I = 1 TO 1000
40      X = 1;  E = RND
50      G = RND;  X = X + 1
60      IF G <= E THEN 50
70      IF X < 11 THEN R (X) = R (X) + 1 ELSE R (1) = R (1) + 1
80      S = S + X
90  NEXT I
100 FOR X = 2 TO 10 PRINT X, R (X)/1000
110 PRINT  " > 10", R (1)/1000
120 PRINT
130 PRINT "MEAN NUMBER CAUGHT  =";S/1000
140 END
```

Fig. 88

X	R(X)/1000
2	0.490
3	0.166
4	0.083
5	0.065
6	0.037
7	0.015
8	0.019
9	0.015
10	0.016
>10	0.094

MEAN NUMBER CAUGHT = 17.126

9. See Fig. 89. X gives the position of the particle, and U counts the returns to the origin.

10. See Fig. 90. M is the current maximum and X gives the position of the particle.

```
10 X = U = 0
20 FOR I = 1 TO 1000
30       X = X + 2 * INT (2 * RND) - 1
40       IF X = 0 THEN U = U + 1
50 NEXT I
60 PRINT U
70 END
```

Fig. 89

```
10 X = M = 0
20 FOR I = 1 TO 1000
30       X = X + 2 * INT (2 * RND) - 1
40       IF ABS (X) > M THEN M = ABS (X)
50 NEXT I
60 PRINT M
70 END
```

Fig. 90

11. a) In Fig. 6.26 lines 40 and 60 should be replaced by

```
40 Z = Z + 2 * INT (2 * RND) - 1
60 IF ABS (Z) < 3 THEN 40
```

b) In Fig. 6.27 lines 80 and 100 should be replaced by

```
80 Z = Z + 2 * INT (2 * RND) - 1
100 IF ABS (Z) < 3 THEN 80
```

6.3

1. $\lambda = 0.9637809877$

2. b) It gives $E(T) = 16$.

d) $\lambda^4 - \lambda^3 + \frac{1}{16} = 0$ has largest positive root $\lambda = 0.9196433776$

3. a) Since $p_{2i} = 0$ we use $\dfrac{p_{2i-1}}{p_{2i+1}}$.

b) $p_{2n} = 0$, $p_{2n+1} = \frac{2}{9}\left(\frac{7}{9}\right)^{n-1}$, $n = 1, 2, 3, \ldots$.

c) See Fig. 91.

d) $q_{2n-1} = q_{2n} = \left(\frac{7}{9}\right)^{n-1}$, $< 10^{-3} \Rightarrow n > 28.48 \Rightarrow n = 29$.

Hence $q_{56} > 10^{-3}$ and $q_{57} < 10^{-3}$.

4. $E(T) = 6\frac{2}{3}$.

5. $E(T) = 25$.

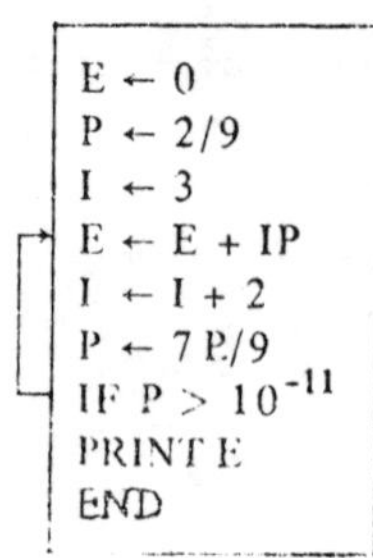

Fig. 91

7.

5. $v_{20} = 20.5 - 2^5 + 1 = 69$. The following sorted sequences are produced successively: ten 2-sequences, five 4-sequences, two 8-sequences, and one 4-sequence, a 12-sequence and an 8-sequence, and a 20-sequence. The number of comparisons is $10 + 15 + 14 + 11 + 19 = 69$. Notice that two sorted sequences of m and n elements can be merged using $m + n - 1$ comparisons.

6. For the algorithms in c) and f) $10^5\,(10^5 - 1)/2 \approx 5 \cdot 10^9$ and 1 568 929 comparisons are needed, respectively.

References

(1) Bauer, F. L. and Weinhart, K.: Informatik (Computer Science), Bayerischer Schulbuch-Verlag, 1974.

(2) Claus, V.: Einführung in die Informatik (Introduction to Computer Science), Teubner, 1975.

(3) Engel, A.: Computerorientierte Mathematik (Computer-oriented Mathematics), MU, April 1975.

(4) Engel, A.: Wahrscheinlichkeitsrechnung und Statistik, Bd. 1 (Probability and statistics, Vol. I), Ernst Klett, 1973.

(5) Engel, A.: Wahrscheinlichkeitsrechnung und Statistik, Bd. 2 (Probability and statistics, Vol. II).

(6) Engel, A.: Anwendungen der Analysis zur Konstruktion mathematischer Modelle, (Using calculus in constructing mathematical models), MU, August 1971.

(7) Forsythe, A., et al.: Computer Science: A First Course, 2nd ed. Wiley, 1975.

(8) Kemeney, J. G. and Kurtz, Th. E.: Basic Programming, 2nd ed. Wiley, 1971.

(9) Klingen, L. et al.: Informatik (Computer Science), Ernst Klett, 1975.

(10) Knuth, D. E.: The Art of Computer Programming, Vols 1, 2, 3, Addison Wesley, 1968-73

(11) Nievergelt, J., et al.: Computer Approaches to Mathematical Problems, Prentice Hall, 1974.

(12) Wirth, N.: Systematic programming, Prentice Hall, 1973

(13) Wirth, N.: Algorithms + data-structures = Programs, Prentice Hall, 1976

(14)(†) Dromey, R. G. : How to solve it by computer, Prentice Hall, 1982.

(15)(†) Engel, A.: The role of algorithms and computers in teaching mathematics at school, IN New trends in mathematics teaching, Vol. IV, UNESCO, 1979.

(Note 7, 8, 10-13 are in English, the others in German.
Refs. 14 and 15 have been added by the translator.)

Index

algorithm 1
area under curve 69, 82-92
Archimedes 69-73, 211
area function inverse hyperbolic function 91
arithmetic - geometric mean 124
arithmetic - harmonic mean 27
assignment operator 1, 2

Bingo 193-4
binomial distribution 187
birthday problem 186
bisection method 109, 126
Brownian movement 203
bubble sort 221
Buffon's needle problem 103-5

collectors problem 194
comparison 221, 224
congruence method 35
continued fraction 64-67
convergence, linear 23, 124
——, quadratic 119-124
—— -factor 23-6, 119-121, 146
——, accelerated 92-7
corrector 153
crap game 195, 217-8
Cusanus 81-2

diffusion process 203
distribution 212
Dukta, J 172

e 86-8, 140-7, 171-2
eight queens problem 227-30
error, absolute 96
——, relative 22, 96
Euclidean algorithm 47-54, 65-7
Euler 102
Euler-polygon 140, 153
—— constant 168
—— method 165
expected value 212
exponent program 16-18, 27-32, 47, 142
exponential function 86-92, 138-147
extrapolation 144

factorial 16
Fibonacci 37-42
fixed point 114-126
——, attracting 117
——, repelling 117
flowchart 7

Gauss, C. F. 97, 98, 124, 226

halving method 109, 126
harmonic series 167-171
harmonic motion 147
Hermite rule 137, 146, 150
Horner rule 44, 45
Huygens, Ch. 91, 96
hyperbolic functions 91

initial value problem 152
instruction 1
integer function 4, 7, 18-21
iteration 114-125

Kommerell, K. 93

Lehmer, D. H. 35
Leibniz series 101
logarithm 27, 32, 33, 66-7
—— , natural 83, 96
Lotto 193-4
loss of accuracy by subtraction 26
lattice point 97-100

mapping, expanding 116
—— , contracting 116-122
maximum program 14-15, 126-9
mid-point rule 129, 135, 154, 166
money changing problem 178
Monte Carlo method 102-7

Neumann, J. von 35, 224
Newton method 118-122, 125
Nicomachus algorithm 50
number theory 43-68

overflow 1

Pascal triangle 173-5
period length 62-4
permutation 179
—— , golden 223
—— , inverse 183
—— , Josephus 185
—— , order of 183
phase space 133
π 70-3, 81-2, 92-105
Picard-iteration 141
planetary motion 160-2
predictor 153
prime numbers 54-62
—— twins 57
program 1
pursuit problems 155-159
Pythagorean triples 51

quadrature, hyperbola 82-92
—— , parabola 69

random digit 29-33
random number 33
— number generator 33-7, 191
— permutation 179, 223
— sample 193
random walk on line 191, 201, 210
— in plane 202-3
— in space 210, 211
— on cube 204, 205, 215
regula falsi 110
records 206
relative frequency 198, 212, 239
Romberg process 132-3
— table 92-6
root function 22-27

secant method 111, 119
simple harmonic motion 147
Simpson's rule 93, 129, 136, 145
sorting 3, 220
sorting by exchanging 220
— by frequency table 222
— by insertion 220
— by merging 224
— by selection 221
— by shellsort 225
standard functions of BASIC 4, 15, 109
state space 133
Stirling, J. 179
subtraction, loss of accuracy by 26

trapezium method 129-132, 150-3, 166
trigonometric functions 74-80, 85-91

uniform distribution 33
underflow 1

Vieta, F. 101

Wallis, J. 101
whole-number function (integer function) 4, 7, 18-21